Roger Penrose's original and provocative ideas about the large-scale phy-
sics of the Universe, the small-scale world of quantum physics and the
physics of the mind have been the subject of controversy and discussion. These ideas were set forth in his best-selling books *The Emperor's
New Mind* and *Shadow of the Mind*. In this book, he summarises and
brings up to date his thinking in these complex areas. He presents a
masterful summary of those areas of physics in which he feels there are
major unsolved problems. Through this, he introduces radically new
concepts which he believes will be fruitful in understanding the work-
ings of the brain and the nature of the human mind. These ideas are then
challenged by three distinguished experts from different backgrounds
– Abner Shimony and Nancy Cartwright as Philosophers of Science and
Stephen Hawking as a Theoretical Physicist and Cosmologist. Finally,
Roger Penrose responds to their thought-provoking criticisms.

This volume provides an accessible, illuminating and stimulating introduction to Roger Penrose's vision of theoretical physics for the twenty-
first century. His enthusiasm, insight and good humour shine through
this brilliant account of the problems of modern physics.

The Large, the Small and the Human Mind

Cambridge University Press gratefully acknowledges the cooperation of the President and Fellows of Clare Hall, Cambridge, under whose auspices the 1995 Tanner Lectures on Human Values (from which this book derives) were held.

The Large, the Small and the Human Mind

ROGER PENROSE

WITH

ABNER SHIMONY

NANCY CARTWRIGHT

AND STEPHEN HAWKING

EDITED BY MALCOLM LONGAIR

CAMBRIDGE
UNIVERSITY PRESS

PUBLISHED BY THE PRESS SYNDICATE OF THE UNIVERSITY OF CAMBRIDGE
The Pitt Building, Trumpington Street, Cambridge

CAMBRIDGE UNIVERSITY PRESS
The Edinburgh Building, Cambridge, United Kingdom
40 West 20th Street, New York, NY 10011-4211, USA
10 Stamford Road, Oakleigh, Melbourne 3166, Australia

© Cambridge University Press 1997

First published 1997
First paperback edition 1999

Printed in the United States of America

Library of Congress Cataloging-in-Publication Data is available.

A catalog record for this book is available from the British Library.

ISBN 0 521 56330 5 hardback
ISBN 0 521 65538 2 paperback

Contents

Notes on the Contributors

ROGER PENROSE is Rouse Ball Professor of Mathematics at the University of Oxford

ABNER SHIMONY is Professor Emeritus of Philosophy and Physics at Boston University

NANCY CARTWRIGHT is Professor of Philosophy, Logic and Scientific Method at the London School of Economics and Political Science

STEPHEN HAWKING is Lucasian Professor of Mathematics at the University of Cambridge

Picture Credits

The Emperor's New Mind, R. Penrose, 1989. Oxford: Oxford University Press. 1.6, 1.8, 1.11, 1.12, 1.13, 1.16(a), (b) and (c), 1.18, 1.19, 1.24, 1.25, 1.26, 1.28(a) and (b), 1.29, 1.30, 2.2, 2.5(a), 3.20.

Shadows of the Mind, R. Penrose, 1994. Oxford: Oxford University Press. 1.14, 2.3, 2.4, 2.5(b), 2.6, 2.7, 2.19, 2.20, 3.7, 3.8, 3.10, 3.11, 3.12, 3.13, 3.14, 3.16, 3.17, 3.18

High Energy Astrophysics, Volume 2, M.S. Longair, 1994. Cambridge: Cambridge University Press. 1.15, 1.22.

Courtesy of Cordon Art-Baarn-Holland ©1989. 1.17, 1.19

Foreword by
Malcolm Longair

One of the more encouraging developments of the last decade has been the publication of a number of books by eminent scientists in which they attempt to communicate the essence and excitement of their science to the lay-reader. Some of the more striking examples include the extraordinary success of Stephen Hawking's *A Brief History of Time*, which is now publishing history, James Gleick's book *Chaos*, showing how successfully an intrinsically difficult subject can be made into a thrilling detective story, and Steven Weinberg's *Dreams of a Final Theory*, which makes the nature and objectives of contemporary particle physics remarkably accessible and compelling.

In this wave of popularisation, Roger Penrose's book *The Emperor's New Mind* of 1989 stood out as quite distinctively different from the others. Whereas other authors aimed to communicate the content and excitement of contemporary science, Roger's book was a strikingly original vision of how many apparently disparate aspects of physics, mathematics, biology, brain science and even philosophy might be subsumed within a new, as yet undefined, theory of fundamental processes. Not surprisingly, *The Emperor's New Mind* stimulated a great deal of controversy and, in 1994, Roger published a second book, *Shadows of the Mind*, in which he attempted to rebut a number of the criticisms of his arguments and offered further insights and developments of his ideas. In his 1995 Tanner lectures, he presented a survey of the central themes discussed in his two books and then participated in a discussion of these with Abner Shimony, Nancy Cartwright and Stephen Hawking. The three lectures reproduced in Chapters 1–3 of this book provide a gentle introduction to the ideas expounded in much more detail in the two books, and the contributions of the three discussants in Chapters 4, 5 and 6 raise many of the concerns which have been expressed about them. Roger has the opportunity to comment upon these concerns in Chapter 7.

Roger's chapters speak eloquently for themselves but a few words of introduction may set the scene for the particular approach he takes to some of the most profound problems of modern science. He has been recognised internationally as one of the most gifted of contemporary mathematicians but his research work has always been placed firmly in a real physical setting. The work for which he is most famous in astrophysics and cosmology concerns theorems in relativistic theories of gravity, some of this work having been carried out jointly with Stephen Hawking. One of the theorems shows that inevitably, according to classical relativistic theories of gravity, within a black hole, there must be a physical singularity, that is, a region of space in which the curvature of space or, equivalently, the density of matter becomes infinitely great. The second states that, according to classical relativistic theories of gravity, there is inevitably a similar physical singularity at the origin of Big Bang cosmological models. These results indicate that, in some sense, there is a serious incompleteness in these theories, since physical singularities should be avoided in all physically meaningful theories.

This is, however, only one aspect of a huge range of contributions to many different areas of mathematics and mathematical physics. The Penrose process is a means by which particles can extract energy from the rotational energy of rotating black holes. Penrose diagrams are used to study the behaviour of matter in the vicinity of black holes. Underlying much of his approach, there is a very strong geometrical, almost pictorial, sense which is present throughout Chapters 1-3. The general public is most familiar with this aspect of his work through the 'impossible' pictures by M. C. Escher and through Penrose tiles. It is intriguing that it was the paper of Roger and his father, L. S. Penrose, which provided the inspiration for a number of Escher's 'impossible' drawings. Furthermore, Escher's Circle Limit pictures are used to illustrate Roger's enthusiasm for hyperbolic geometries in Chapter 1. Penrose tilings are remarkable geometrical constructions in which an infinite plane can be completely tiled by tiles of a small number of different shapes. The most amazing examples of these tilings are those which can completely cover an infinite plane but which are non-repeating – in other words, the same pattern of tiles does not repeat itself at any point on the infinite plane. This theme recurs in Chapter 3 in connection with the issue of whether or not specific sets of precisely defined mathematical procedures can be carried out by computer.

Roger thus brings a formidable array of mathematical weapons as well as an extraordinary range of achievement in mathematics and physics to some of the most profound problems of modern physics. There is no question about the reality and importance of the problems he addresses. Cosmologists have good reasons to be firmly convinced that the Big Bang provides the most convincing picture we have for understanding the large-scale features of our Universe. It is, however, seriously incomplete in a number ways. Most cosmologists are convinced that we have a good understanding of the basic physics needed to account for the overall properties of the Universe from about the time it was a thousandth of a second old to the present day. The picture only comes out right, however, if we arrange the initial conditions rather carefully. The big problem is that we run out of tried and tested physics when the Universe was significantly less than a thousandth of a second old and so we have to rely upon reasonable extrapolations of the known laws of physics. We know rather well what these initial conditions must have been, but why they came about is a matter of speculation. There is general agreement that these are among the most important problems of contemporary cosmology.

A standard framework has been developed for attempting to resolve these problems, known as the inflationary picture of the early Universe. Even in this picture, certain features of our Universe are assumed to originate at the very earliest meaningful times, at what is known as the Planck epoch, when it becomes necessary to understand quantum gravity. This epoch occurred when the Universe was only about 10^{-43} seconds old, which may seem somewhat extreme but, on the basis of what we know today, we have to take seriously what happened at these very extreme epochs.

Roger accepts the conventional Big Bang picture, so far as it goes, but he rejects the inflationary picture of its early stages. Rather, he believes that there is some missing physics which must be associated with a proper quantum theory of gravity, a theory which we do not yet possess, despite the fact that theorists have been trying to solve that problem for many years. Roger argues that they have been trying to solve the wrong problem. Part of his concern relates to the problem of the entropy of the Universe as a whole. Since entropy, or to put it more simply, disorder, increases with time, the Universe must have started in a highly ordered state of very small entropy indeed. The probability of this coming about by chance is vanishingly small. Roger argues that this problem should be solved as part of the correct theory of quantum gravity.

The necessity of quantisation leads to his discussion in Chapter 2 of the problems of quantum physics. Quantum mechanics, and its relativistic extension in quantum field theory, have been phenomenally successful in accounting for many experimental results in particle physics and in the properties of atoms and particles. It took many years, however, before the full physical significance of the theory was appreciated. As Roger illustrates beautifully, the theory contains as part of its intrinsic structure highly non-intuitive features, which have no counterpart in classical physics. For example, the phenomenon of non-locality means that, when a matter–antimatter pair of particles is produced, each particle maintains a 'memory' of the creation process, in the sense that they cannot be considered to be completely independent of each other. As Roger expresses it, 'Quantum entanglement is a very strange thing. It is somewhere between objects being separate and being in communication with each other'. Quantum mechanics also allows us to obtain information about processes which could have happened but didn't. The most striking example he discusses is the amazing Elitzur–Vaidman bomb-testing problem which illustrates just how different quantum mechanics is from classical physics.

These non-intuitive features are part of the structure of quantum physics but there are deeper problems. Those which Roger focuses upon concern the way in which we relate phenomema occurring at the quantum level to the macroscopic level of making an observation of a quantum system. This is a controversial area. Most practising physicists simply use the rules of quantum mechanics as computational tools which happen to give extraordinarily accurate answers. If we apply the rules correctly, we will get the correct answers. This involves, however, a somewhat inelegant process for translating phenomena from the simple linear world at the quantum level to the world of real experiment. This process involves what is known as the 'collapse of the wavefunction' or the 'reduction of the state vector'. Roger believes that some fundamental pieces of physics are missing from the conventional picture of quantum mechanics. He argues that a completely new theory is needed which incorporates what he calls the 'objective reduction of the wavefunction' as an integal part of the theory. This new theory must reduce to conventional quantum mechanics and quantum field theory in the appropriate limit but it is likely to bring with it new physical phenomena. In these, there may lie solutions to the problem of quantising gravity and the physics of the early Universe.

In Chapter 3, Roger seeks to unearth common features between mathematics, physics and the human mind. It often comes as a surprise that the most rigorously logical of the sciences, abstract mathematics, cannot be programmed on a digital computer, no matter how accurate it is and how large its memory. Such a computer cannot discover mathematical theorems in the way that human mathematicians do. This surprising conclusion is derived from a variant of what is called Gödel's Theorem. Roger interprets this to mean that the processes of mathematical thinking, and by extension all thinking and conscious behaviour, are carried out by 'non-computational' means. This is a very fruitful clue because our intuition tells us that the huge variety of our conscious perceptions are also 'non-computational'. Because of the central importance of this result for his general argument, he devoted over half of *Shadows of the Mind* to showing that his interpretation of Gödel's Theorem was watertight.

Roger's vision is that, in some way, the problems of quantum mechanics and the problems of understanding consciousness are related in a number of ways. Non-locality and quantum coherence suggest, in principle, ways in which large areas of the brain could act coherently. The non-computational aspects of consciousness he believes may be related to the non-computational processes which may be involved in the objective reduction of the wave-function to macroscopic observables. Not content with simply enunciating general principles, he attempts to identify the types of structure within the brain which could be capable of sustaining such types of new physical process.

This summary does scant justice to the originality and fecundity of these ideas and the brilliance with which they are developed in this book. Throughout the exposition, several underlying themes play an important role in determining the direction of his thought. Perhaps the most important is the remarkable ability of mathematics to describe fundamental processes in the natural world. As Roger expresses it, the physical world in a sense emerges from the Platonic world of mathematics. But, we do not derive new mathematics from the need to describe the world, or from making experiment and observation fit mathematical rules. Understanding of the structure of the world can come from broad general principles and the mathematics itself.

It is scarcely surprising that these bold proposals have been the subject of controversy. A flavour of many of the concerns voiced by experts coming from very different intellectual backgrounds is provided by the contributions of the disputants. Abner Shimony agrees with Roger about a number of his objectives – he agrees that there is some incompleteness in the standard formulation of quantum mechanics, along the same lines outlined by Roger and he agrees that quantum mechanical concepts are relevant to the understanding of the human mind. He claims, however, that Roger 'is an alpinist who has tried to climb the wrong mountain' and suggests alterative ways of looking at the same areas of concern in a constructive way. Nancy Cartwright raises the basic question of whether or not physics is the correct starting point for understanding the nature of consciousness. She also raises the thorny problem of how the laws which govern different scientific disciplines can actually be derived from one another. Most critical of all is Stephen Hawking, Roger's old friend and colleague. In many ways, Hawking's position is the closest to what might be called the standard position of the 'average' physicist. He challenges Roger to develop a detailed theory of the objective reduction of the wavefunction. He denies that physics has much of value to say about the problem of consciousness. These are all justifiable concerns but Roger defends his position in his response to the disputants in the final chapter of this book.

What Roger has succeeded in doing is in creating a vision or manifesto for how mathematical physics might develop in the twenty-first century. Through Chapters 1–3, he builds up a connected narrative which suggests how each part of the story might fit into a coherent picture of a completely new type of physics which has built into it his central concerns of non-computability and the objective reduction of the wavefunction. The test of these concepts will depend upon the ability of Roger and others to bring into being a realisation of this new type of physical theory. And, even if this programme were not to be immediately successful, are the ideas inherent in the general concept fruitful for the future development of theoretical physics and mathematics? It would be very surprising indeed if the answer were 'No'.

CHAPTER 1

Space-time and Cosmology

The title of this book is *The Large, the Small and the Human Mind* and the subject of this first chapter is the Large. The first and second chapters are concerned with our physical Universe, which I represent very schematically as the 'sphere' in Figure 1.1. However, these will not be 'botanical' chapters, telling you in detail what is here and what is there in our Universe, but rather I want to concentrate upon understanding of the actual laws which govern the way the world behaves. One of the reasons that I have chosen to divide my descriptions of the physical laws between two chapters, namely, the Large and the Small, is that the laws which govern the large-scale behaviour of the world and those which govern its small-scale behaviour seem to be very different. The fact that they seem to be so different, and what we might have to do about this seeming discrepancy, is central to the subject of the Chapter 3 – which is where the human mind comes in.

Since I shall be talking about the physical world in terms of the physical theories which underlie its behaviour, I shall also have to say something about another world, the Platonic world of absolutes, in its particular role as the world of mathematical truth. One can well take the view that the 'Platonic world' contains other absolutes, such as the Good and the Beautiful, but I shall be concerned here only with the Platonic concepts of mathematics.

1

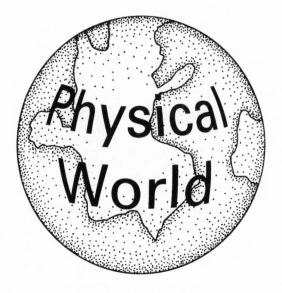

Fig. 1.1.

Some people find it hard to conceive of this world as existing on its own. They may prefer to think of mathematical concepts merely as idealisations of our physical world – and, on this view, the mathematical world would be thought of as emerging from the world of physical objects (Figure 1.2).

Now, this is not how I think of mathematics, nor, I believe, is it how most mathematicians or mathematical physicists think about the world. They think about it in a rather different way, as a structure precisely governed according to timeless mathematical laws. Thus, they prefer to think of the physical world, more appropriately, as emerging out of the ('timeless') world of mathematics, as illustrated in Figure 1.3. This picture will have importance for what I shall say in the Chapter 3, and it also underlies most of what I shall say in the Chapters 1 and 2.

2

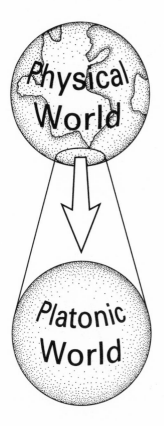

Fig. 1.2.

One of the remarkable things about the behaviour of the world is how it seems to be grounded in mathematics to a quite extraordinary degree of accuracy. The more we understand about the physical world, and the deeper we probe into the laws of nature, the more it seems as though the physical world almost evaporates and we are left only with mathematics. The deeper we understand the laws of physics, the more we are driven into this world of mathematics and of mathematical concepts.

3

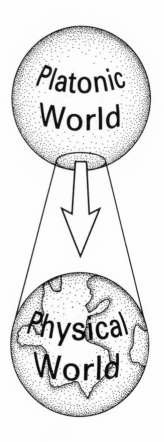

Fig. 1.3.

Let us look at the scales we have to deal with in the Universe and also the role of our place in the Universe. I can summarise all these scales in a single diagram (Figure 1.4). On the left-hand side of the diagram, time-scales are shown and, on the right-hand side, are the corresponding distance-scales. At the bottom of the diagram, on the left-hand side, is the very shortest time-scale which is physically meaningful. This time-scale is about 10^{-43} of a sec-

4

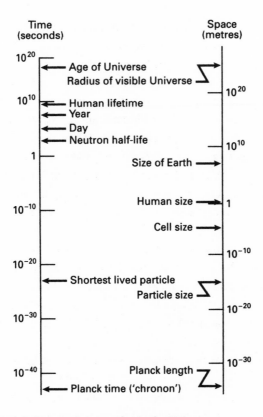

Fig. 1.4. Sizes and time-scales in the Universe.

ond and is often referred to as the *Planck time-scale* or a 'chronon'. This time-scale is much shorter than anything experienced in particle physics. For example, the shortest lived particles, called resonances, last for about 10^{-23} of a second. Further up the diagram, on the left, the day and the year are shown, and at the top of the diagram, the present age of the Universe is shown.

On the right-hand side of the diagram, distances corresponding to these time-scales are depicted. The length corresponding to the Planck time (or chronon), is the fundamental unit of length, called the *Planck length*. These concepts of the Planck time and the Planck length fall out naturally when one tries to combine the physical theories which describe the large and the small, that is, combining Einstein's General Relativity, which describes the physics of the very large, with quantum mechanics, which describes the physics of the very small. When these theories are brought together, these Planck lengths and times turn out to be fundamental. The translation from the left-hand to the right-hand axis of the diagram is via the speed of light so that times can be translated into distances by asking how far a light signal could travel in that time.

The sizes of the physical objects represented on the diagram range from about 10^{-15} of a metre for the characteristic size of particles to about 10^{27} metres for the radius of the observable Universe at the present time, which is roughly the age of the Universe multiplied by the speed of light. It is intriguing to note where *we* are in the diagram, namely the human scale. With regard to spatial dimensions, it can be seen that we are more of less in the middle of the diagram. We are enormous compared with the Planck length; even compared with the size of particles, we are very large. Yet, compared with the distance scale of the observable Universe, we are very tiny. Indeed, we are much smaller compared with it than we are large compared with particles. On the other hand, with regard to temporal dimensions, the human lifetime is almost as long as the Universe! People talk about the ephemeral nature of existence but, when you look at the human lifetime as shown in the diagram, it can be seen that we are not ephemeral at all – we live more or less as long as the Universe itself! Of course, this is looking on a 'logarithmic scale', but this is the natural thing to do when we are concerned with such enormous ranges. To put it another way, the number of human lifetimes which make up the age

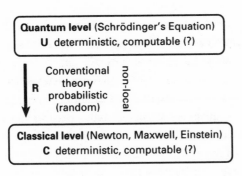

Fig. 1.5.

of the Universe is very, very much less than the number of Planck times, or even lifetimes of the shortest lived particles, which make up a human lifetime. Thus, we are really very stable structures in the Universe. As far as spatial sizes are concerned, we are very much in the middle – we directly experience neither the physics of the very large nor the very small. We are very much in-between. In fact, looked at logarithmically, all living objects from single cells to trees and whales are roughly the same in-between size.

What kinds of physics apply on these different scales? Let me introduce the diagram which summarises the whole of physics (Figure 1.5). I have had to leave out a few details, of course, such as all the equations! But the essential basic theories that physicists use are indicated.

The key point is that, in physics, we use two very different types of procedure. To describe the small-scale behaviour, we use quantum mechanics – what I have described as the quantum level in Figure 1.5. I shall say much more about this in the Chapter 2. One of the things which people say about quantum mechanics is that it is fuzzy and indeterministic, but this is not true. So long as you remain at this level, quantum theory is deterministic and pre-

7

cise. In its most familiar form, quantum mechanics involves use of the equation known as Schrödinger's Equation which governs the behaviour of the physical state of a quantum system – called its *quantum state* – and this is a deterministic equation. I have used the letter U to describe this quantum level activity. Indeterminacy in quantum mechanics only arises when you perform what is called 'making a measurement' and that involves magnifying an event from the quantum level to the classical level. I shall say quite a lot about this in the Chapter 2.

On the large scale, we use classical physics, which is entirely deterministic – these classical laws include Newton's laws of motion, Maxwell's laws for the electromagnetic field, which incorporate electricity, magnetism and light, and Einstein's theories of relativity, the Special Theory which deals with large velocities and the General Theory which deals with large gravitational fields. These laws apply very, very accurately on the large scale.

Just as a footnote to Figure 1.5, it can be seen that I have included a remark about 'computability' in quantum and classical physics. This has no relevance to this chapter or Chapter 2, but it will have importance in Chapter 3, and I shall return to the issue of computability there.

For the rest of the present chapter, I shall be primarily concerned with Einstein's theory of relativity – specifically, how the theory works, its extraordinary accuracy and something about its elegance as a physical theory. But let us first consider Newtonian theory. Newtonian physics, just as in the case of relativity, allows a space-time description to be used. This was first precisely formulated by Cartan for Newtonian gravity, some time after Einstein had presented his General Theory of relativity. The physics of Galileo and Newton is represented in a space-time for which there is a global time coordinate, here depicted as running up the diagram (Figure 1.6); and for each constant value of the

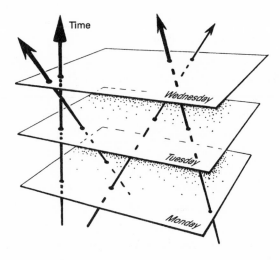

Fig. 1.6. Galilean space-time: particles in uniform motion are depicted as straight lines.

time, there is a space section which is a Euclidean 3-space, here depicted as horizontal planes. An essential feature of the Newtonian space-time picture is that these space-slices, across the diagram, represent moments of simultaneity.

Thus, everything which occurs on Monday at noon lies on one horizontal slice through the space-time diagram; everything which happens on Tuesday at noon lies on the next slice shown in the diagram and so on. Time cuts across the space-time diagram and the Euclidean sections follow one after the other as time progresses. All observers, no matter how they move through the space-time, can agree about the time when events take place because everyone uses the same time-slices to measure how time passes.

In Einstein's Special Theory of relativity, one has to adopt a different picture. In it, the space-time picture is absolutely essential

9

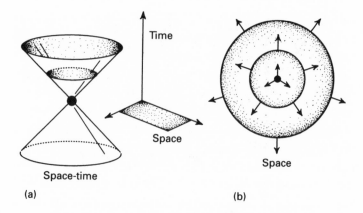

Fig. 1.7. The representation of the history of a light flash
in terms of its propagation in (a) space-time, and (b) space.

– the key difference is that time is not the universal thing it is in
Newtonian theory. To appreciate how the theories differ, it is necessary to understand an essential part of relativity theory, namely
those structures known as *light cones*.

What is a light cone? A light cone is drawn in Figure 1.7. We
imagine a flash of light taking place at some point at some instant – that is, at an *event* in space-time – and the light waves
travel outwards from this event, the source of the flash, at the
speed of light. In a purely spatial picture (Figure 1.7(b)), we can
represent the paths of the light waves through space as a sphere
expanding at the speed of light. We can now translate this motion of the light waves into a space-time diagram (Figure 1.7(a))
in which time runs up the diagram and the space coordinates
refer to horizontal displacements, just as in the Newtonian situation of Figure 1.6. Unfortunately, in the full space-time picture,
Figure 1.7(a), we can only represent two spatial dimensions horizontally on the diagram, because the space-time of our picture is

only three-dimensional. Now, we see that the flash is represented by a point (event) at the origin and that the subsequent paths of the light rays (waves) cut the horizontal 'space' planes in circles, the radii of which increase at the speed of light up the diagram. It can be seen that the paths of the light rays form cones in the space-time diagram. The light cone thus represents the history of this flash of light – light propagates away from the origin along the light cone, which means at the speed of light, into the future. Light rays can also arrive at the origin along the light cone from the past – that part of the light cone is known as the past light cone and all information carried to the observer by light waves arrives at the origin along this cone.

Light cones represent the most important structures in space-time. In particular, they represent the limits of causal influence. The history of a particle in space-time is represented by a line travelling up the space-time diagram, and this line has to lie within the light cone (Figure 1.8). This is just another way of saying that a material particle cannot travel faster than the speed of light. No signal can travel from inside to outside the future light cone and so the light cone does indeed represent the limits of causality.

There are some remarkable geometrical properties that relate to the light cones. Let us consider two observers moving at different speeds through space-time. Unlike the case of Newtonian theory, in which the planes of simultaneity are the same for all observers, there is no absolute simultaneity in relativity. Observers moving at different speeds draw their own planes of simultaneity as different sections through space-time, as illustrated in Figure 1.9. There is a very well-defined way of transforming from one plane to another through what is known as a *Lorentz transformation*, these transformations constituting what is called the *Lorentz group*. The discovery of this group was an essential in-

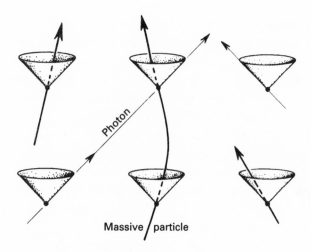

Fig. 1.8. Illustrating the motion of a particle in the
space-time of Special Relativity which is known as
Minkowski space-time or Minkowski geometry. The light
cones at different points in space-time are lined up and
particles can only travel within their future light cones.

gredient in the discovery of Einstein's Special Theory of relativity.
The Lorentz group can be understood as a group of (linear) space-
time transformations, leaving a light cone invariant.

We can also appreciate the Lorentz group from a slightly different viewpoint. As I have emphasised, the light cones are the
fundamental structures of space-time. Imagine that you are an
observer located somewhere in space, looking out at the Universe.
What you see are the light rays coming from the stars to your
eyes. According to the space-time viewpoint, the events you observe are the intersections of the world-lines of the stars with your
past light cone, as illustrated in Figure 1.10(a). You observe along
your past light cone the positions of the stars at particular points.
These points seem to be situated on the celestial sphere that ap-

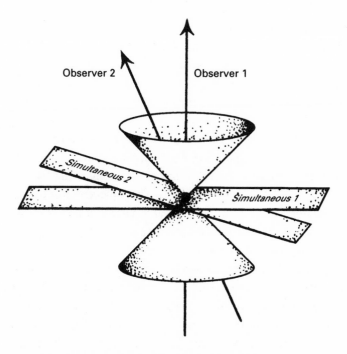

Fig. 1.9. Illustrating the relativity of simultaneity
according to Einstein's Special Theory of relativity.
Observers 1 and 2 are moving relative to one another
through space-time. Events which are simultaneous for
Observer 1 are not simultaneous for Observer 2 and *vice
versa*.

pears to surround you. Now, imagine another observer, moving
at some great speed relative to you, who passes closely by you at
the moment you both look out at the sky. This second observer
perceives the same stars as you do, but finds them to be located
in different positions on the celestial sphere (Figure 1.10(b)) – this
is the effect known as *aberration*. There is a set of transformations which enables us to work out the relationship between what

each of these observers sees on his or her celestial sphere. Each of these transformations is one which takes a sphere to a sphere. But it is one of a very special kind. It takes exact circles to exact circles and it preserves angles. Thus, if a pattern in the sky appears to be circular to you, then it must appear circular also to the other observer.

There is a very beautiful way of describing how this works and I illustrate it to show that there is a particular elegance in the mathematics which often underlies physics at its most fundamental level. Figure 1.10(c) shows a sphere with a plane drawn through its equator. We can draw figures on the surface of the sphere and then examine how they are projected to the equatorial plane from the south pole, as illustrated. This type of projection is known as a stereographic projection and it has some rather extraordinary properties. Circles on the sphere are projected into exact circles on the plane, and the angles between curves on the sphere are projected into exactly the same angles on the plane. As I shall discuss more fully in the Chapter 2 (cf. Figure 2.4), this projection allows us to label the points of the sphere by complex numbers (numbers involving the square root of -1), numbers which are also being used to label the points of the equatorial plane, together with 'infinity', to give the sphere the structure known as the 'Riemann sphere'.

For those who are interested, the aberration transformation is

$$u \to u' = \frac{\alpha u + \beta}{\gamma u + \delta}$$

and, as is well known to mathematicians, this transformation sends circles into circles and preserves angles. Transformations of this kind are known as Möbius transformations. For our present purposes, we need merely note the simple elegance of the form

14

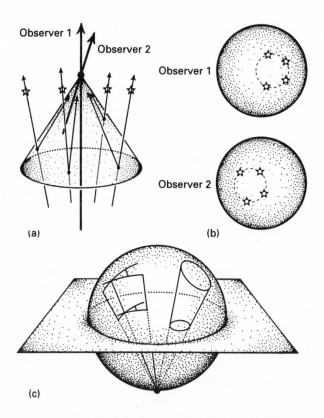

Fig. 1.10. Illustrating how observations are made of the sky by Observers 1 and 2. (a) Observers 1 and 2 observe stars along the past light cone. The points at which stars cross the light cone are indicated by black dots. Light signals propagate from the stars to the observers along the light cone as illustrated. Observer 2 is moving through space-time at a certain speed relative to Observer 1. (b) Illustrating the location of stars on the sky as observed by Observer 1 and Observer 2, when they are coincident at some point in space-time. (c) A good way to represent the transformation of the sky between the two observers is via stereographic projection: circles map to circles, and angles are preserved.

of the Lorentz (aberration) formula when written in terms of such a complex parameter u.

A striking point about this way of looking at these transformations is that, according to Special Relativity, the formula is very simple, whereas, in expressing the corresponding aberration transformation according to Newtonian Mechanics, the formula would be much more complicated. It often turns out that, when you get down to the fundamentals and develop a more exact theory, the mathemetics turns out to be simpler, even if the formalism appears to be more complicated in the first instance. This important point is exempified by the contrast between Galilean and Einsteinian relativity.

Thus, in the Special Theory of relativity, we have a theory which is, in many ways, simpler than Newtonian mechanics. From the point of view of mathematics, and particularly from the point of view of group theory, it is a much nicer structure. In Special Relativity, the space-time is flat and all the light cones are lined up regularly as illustrated in Figure 1.8. If we now go one step further to Einstein's General Relativity, that is, the theory of space-time in the presence of gravity, the picture seems at first sight rather muddied up – the light cones are all over the place (Figure 1.11). Now, I have been saying that, as we develop deeper and deeper theories, the mathematics becomes simpler, but look what has happened here – I had a nice elegant piece of mathematics which has become horribly complicated. Well, that sort of thing happens – you will have to bear with me for a little while until the simplicity reappears.

Let me remind you of the fundamental ingredients of Einstein's theory of gravity. One basic ingredient is called Galileo's Principle of Equivalence. In Figure 1.12(a), I show Galileo leaning over from the top of the Tower of Pisa dropping large and small rocks. Whether or not he actually performed this experiment, he cer-

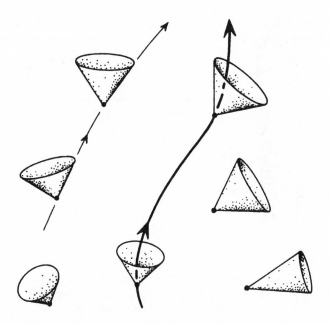

Fig. 1.11. A picture of curved space-time.

tainly well understood that, if the effects of air resistance are ignored, the two rocks would fall to the ground in the same time.
If you happened to be sitting on one of these rocks looking at the
other one as they fall together, you would observe the other rock
hovering in front of you (I have shown a camcorder attached to
one of the rocks to make the observation). Nowadays, with space
travel, this is a very familiar phenomenon – just recently, we have
seen a British-born astronaut walking in space, and, just like the
big rock and the little rock, the spaceship hovers in front of the
astronaut – this is exactly the same phenomenon as Galileo's Principle of Equivalence.

Thus, if you look at gravity in the right way, that is, in a falling
frame of reference, it seems to disappear right in front of your

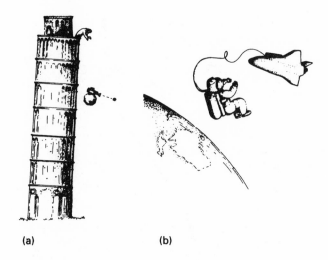

(a)   (b)

Fig. 1.12. (a) Galileo dropping two rocks (and a camcorder)
from the Leaning Tower of Pisa. (b) The astronaut sees the
space-vehicle hover before him or her, seemingly
unaffected by gravity.

eyes. This is indeed correct. But Einstein's theory does *not* tell
you that gravity disappears – it only tells you that the *force* of
gravity disappears. There is something left and that is the tidal
effect of gravity.

Let me introduce a little bit more mathemetics, but not much.
We need to describe the curvature of space-time and this is described by an object known as a *tensor* which I have called **Riemann** in the following equation. It is actually called the Riemann
curvature tensor but I will not tell you what it is except that it is
represented by a capital **R** with a number of indices stuck on the
bottom, which are indicated by the dots. The Riemann curvature
tensor is made up of two pieces. One of the pieces is called the

Weyl curvature and the other piece is called the **Ricci** curvature, and we have the (schematic) equation

$$\textbf{Riemann} = \textbf{Weyl} + \textbf{Ricci}$$

$$R_{\ldots} = C_{\ldots} + R_{.}g_{..}$$

Formally, $C_{\ldots}$ and $R_{..}$ are the Weyl and Ricci curvature tensors, respectively, and $g_{..}$ is the metric tensor.

The Weyl curvature effectively measures the tidal effect. What is the 'tidal' effect? Recall that, from the astronaut's point of view, it seems that gravity has been abolished but that is not quite true. Imagine that the astronaut is surrounded by a sphere of particles, which are initially at rest with respect to the astronaut. Now, initially they will just hover there but soon they will start to accelerate because of the slight differences in the gravitational attraction of the Earth at different points on the sphere. (Notice that I am describing the effect in Newtonian language, but that is quite adequate.) These slight differences cause the original sphere of particles to become distorted into an elliptical arrangement, as illustrated in Figure 1.13(a).

This distortion occurs partly because of the slightly greater attraction of the Earth for those particles closer to the Earth and the lesser attraction for those further away, and partly because, at the sides of the sphere, the Earth's attraction acts slightly inwards. This causes the sphere to be distorted into an ellipsoid. It is called the tidal effect for the very good reason that if you replace the Earth by the Moon and the sphere of particles by the Earth with its oceans, then the Moon has the same gravitational effect on the surface of the oceans as the Earth does on the sphere of particles – the sea surface closest to the Moon is pulled towards it, whereas that on the other side of the Earth is, in effect, pushed away from it. The effect causes the sea surface to bulge

19

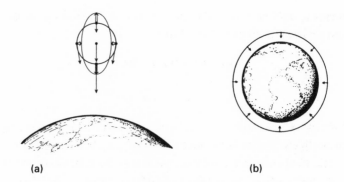

(a) (b)

Fig. 1.13. (a) The tidal effect. Double arrows show relative acceleration. (b) When the sphere surrounds matter (here on Earth), there is a net inward acceleration.

out on either side of the Earth and is the cause of the two high tides which occur each day.

The effects of gravity, from Einstein's point of view, are simply this tidal effect. It is defined essentially by the Weyl curvature, that is, the part $C_{\cdots}$ of the Riemann curvature. This part of the curvature tensor is volume-preserving – that is, if you work out the initial accelerations of the particles of the sphere, the volume of the sphere and the volume of the ellipsoid into which it is distorted are initially the same.

The remaining part of the curvature is known as the *Ricci* curvature and it has a volume-reducing effect. From Figure 1.13(b), it can be seen that, if instead of being at the bottom of the diagram, the Earth were inside the sphere of particles, the volume of the sphere of particles would be reduced as the particles accelerate inwards. The amount of this reduction in volume is a measure of the Ricci curvature. Einstein's theory tells us that the Ricci curvature is determined by the amount of matter present within a small sphere about that point in space. In other words, the density of matter, appropriately defined, tells us how the particles are accelerated inwards at that point in space. Einstein's theory is almost the same as Newton's when expressed in this way.

This is how Einstein formulated his theory of gravity – it is expressed in terms of the tidal effects which are measurements of the local space-time curvature. It is crucial that we have to think in terms of the curvature of four-dimensional space-time. This was shown schematically in Figure 1.11 – we think of the lines which represent the world-lines of particles and the ways in which these paths are distorted as a measurement of the curvature of space-time. Thus, Einstein's theory is essentially a geometric theory of four-dimensional space-time – it is an extraordinarily beautiful theory mathematically.

The history of Einstein's discovery of the theory of General Relativity contains an important moral. It was first fully formulated in 1915. It was not motivated by any observational need but by various aesthetic, geometric and physical desiderata. The key ingredients were Galileo's Principle of Equivalence, exemplified by his dropping rocks of different masses (Figure 1.12), and the ideas of non-Euclidean geometry, which is the natural language for describing the curvature of space-time. There was not a great deal on the observational side in 1915. Once General Relativity was formulated in its final form, it was realised that there were three key observational tests of the theory. The perihelion of the orbit of Mercury advances, or is swung around, in a way which could not be explained by the Newtonian gravitational influence of the other planets – General Relativity predicts exactly the observed advance. The paths of light rays are bent by the Sun and this was the reason for the famous eclipse expedition of 1919, led by Arthur Eddington, which found a result consistent with Einstein's prediction (Figure 1.14(a)). The third test was the prediction that clocks run slow in a gravitational potential – that is, a clock closer

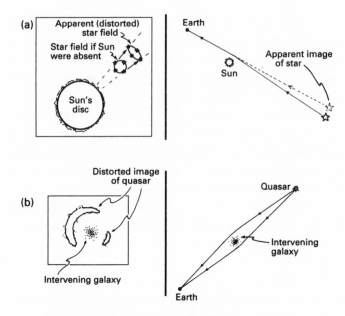

Fig. 1.14. (a) Direct observational effects of gravity on light according to General Relativity. The *Weyl* space-time curvature manifests itself as a distortion of the distant star field, here owing to the light-bending effect of the Sun's gravitational field. A circular pattern of stars would be distorted into an elliptical one. (b) Einstein's light-bending effect is now an important tool in observational astronomy. The mass of the intervening galaxy may be estimated by how much it distorts the image of a distant quasar.

to the ground runs slow with respect to a clock at the top of a tower. This effect has also been measured experimentally. These were never, however, very impressive tests – the effects were always very small and various different theories could have given the same results.

The situation has now changed dramatically – in 1993, Hulse and Taylor were awarded the Nobel prize for a most extraordinary series of observations. Figure 1.15(a) shows the binary pulsar known as PSR 1913+16 – it consists of a pair of neutron stars, each of which is an enormously dense star which has mass about that of the Sun but is only a few kilometres in diameter. The neutron stars orbit about their common centre of gravity in highly elliptical orbits. One of them has a very strong magnetic field and particles are swung round and emit intense radiation which travels to the Earth, some 30 000 light years away, where it is observed as a series of well-defined pulses. All sorts of very precise observations have been made of the arrival times of these pulses. In particular, all the properties of the orbits of the two neutron stars can be worked out as well as all the tiny corrections due to General Relativity.

There is, in addition, a feature which is completely unique to General Relativity, and not present at all in the Newtonian theory of gravity. That is that objects in orbit about each other radiate energy in the form of gravitational waves. These are like light waves but are ripples in space-time rather than ripples in the electromagnetic field. These waves take energy away from the system at a rate which can be precisely calculated according to Einstein's theory and the rate of loss of energy of the binary neutron star system agrees very precisely with the observations, as illustrated by Figure 1.15(b), which shows the speed-up of the orbital period of the neutron stars, measured over 20 years of observation. These signals can be timed so accurately that, over 20 years, the accuracy with which the theory is known to be correct amounts to about one part in 10^{14}. This makes General Relativity the most accurately tested theory known to science.

There is a moral in this story – Einstein's motivations for devoting eight or more years of his life to deriving the General Theory

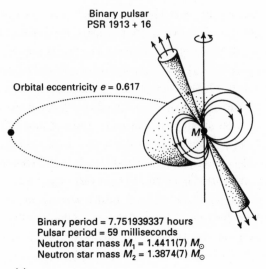

Binary pulsar
PSR 1913 + 16

Orbital eccentricity $e = 0.617$

Binary period = 7.751939337 hours
Pulsar period = 59 milliseconds
Neutron star mass $M_1 = 1.4411(7)$ $M_\odot$
Neutron star mass $M_2 = 1.3874(7)$ $M_\odot$

(a)

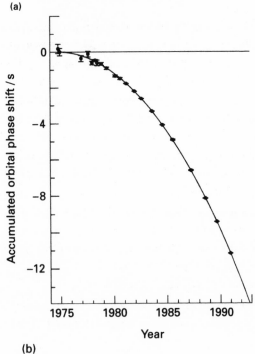

(b)

were not observational or experimental. Sometimes people argue that, 'Well, physicists look for patterns in their experimental results and then they find some nice theory which agrees with these. Maybe this explains why mathematics and physics work so well together.' But, in this case, things were not like that at all. The theory was developed originally without any observational motivation – the mathematical theory is very elegant and it is physically very well motivated. The point is that the mathematical structure is just there in Nature, the theory really is out there in space – it has not been imposed upon Nature by anyone. That is one of the essential points of this chapter. Einstein revealed something that was there. Moreover, it was not just some minor piece of physics he discovered – it is the most fundamental thing that we have in Nature, the nature of space and time.

Here is a very clear case – it goes back to my original diagram concerning the relation between the world of mathematics and

Fig. 1.15. (a) A schematic representation of the binary pulsar PSR 1913 + 16. One of the neutron stars is a radio pulsar. Radio emission is emitted along the poles of the magnetic dipole which is misaligned with respect to the rotation axis of the neutron star. Sharply defined pulses are observed when the narrow beam of radiation is swept across the line of sight to the observer. The properties of the two neutron stars have been derived from very precise timing of the arrival times of the pulses using (and verifying) effects which are only present in Einstein's General Relativity. (b) The change of phase of the arrival times of the pulses from the binary pulsar PSR 1913 + 16, compared with the expected change due to the emission of gravitational radiation by the binary neutron star system (solid line).

the physical world (Figure 1.3). In General Relativity, we have some kind of structure which really does underlie the behaviour of the physical world in an extraordinarily precise way. The way in which these fundamental features of our world are discovered is often not by looking at the way in which Nature behaves, although that is obviously very important. One has to be prepared to throw out theories which might appeal for all sorts of other reasons but which do not fit the facts. But here we have a theory which does fit the facts with extraordinary accuracy. The accuracy involved is about twice as many figures as one has in Newtonian theory, in other words, General Relativity is known to be correct to one part in 10^{14} whereas Newtonian theory is only accurate to one part in 10^7. The improvement is like the increase in the accuracy with which Newton's theory was known to be correct between the seventeenth century and now. Newton knew his theory was correct to about one part in 1000, whereas now it is known to be accurate to one part in 10^7.

Einstein's General Relativity is just a theory, of course. What about the structure of the actual world? I said this chapter would not be botanical but, if I talk about the Universe as a whole, that is not being botanical, since I shall consider only the one Universe as a whole that is given to us. There are three types of standard model which come out of Einstein's theory and these are defined by one parameter, which is, in effect, the one denoted by k in Figure 1.16. There is another parameter which sometimes appears in cosmological arguments which is known as the cosmological constant. Einstein regarded his introduction of the cosmological constant into his equations of General Relativity as his greatest mistake and so I shall leave it out too. If we are forced to bring it back, well, we shall have to live with it.

Assuming the cosmological constant is zero, the three types of universe, which are described by the constant k, are illustrated in

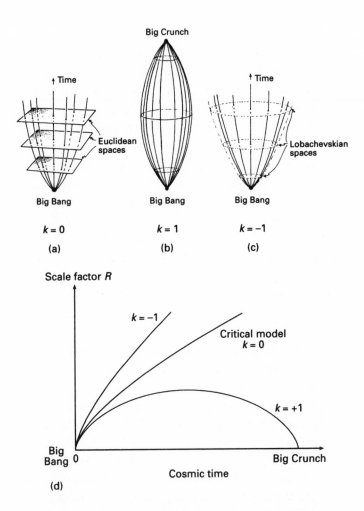

Fig. 1.16. (a) Space-time picture of an expanding universe with Euclidean spatial sections (two space dimensions depicted): $k = 0$. (b) As in (a), but for an expanding (and subsequently contracting) universe with spherical spatial sections: $k = +1$. (c) As in (a) but for an expanding universe with Lobachevskian spatial sections: $k = -1$. (d) The dynamics of the three different types of Friedman model.

27

Figure 1.16. In the diagrams, k takes values 1, 0 and -1, because all the other properties of the models have been scaled away. A better way would have been to talk about the age or scale of the Universe, and then one would have a continuous parameter but, qualitatively, the three different models can be thought of as being defined by the curvature of the space sections of the Universe. If the space sections of Universe are flat, they have zero curvature and $k = 0$ (Figure 1.16(a)). If the space sections are positively curved, meaning that the Universe closes in on itself, then $k = +1$ (Figure 1.16(b)). In all these models the Universe has an initial singular state, the Big Bang, which marks the beginning of the Universe. But in the $k = +1$ case, it expands to a maximum size and then recollapses to a Big Crunch. Alternatively, there is the $k = -1$ case, in which the Universe expands forever (Figure 1.16(c)). The $k = 0$ case is the limiting boundary between the $k = 1$ and $k = -1$ cases. I have shown the radius–time relations for these three types of universe in Figure 1.16(d). The radius can be thought of as some typical scale in the Universe and it can be seen that only the case $k = +1$ collapses to a Big Crunch, while the other two expand indefinitely.

I want to consider the $k = -1$ case in a little more detail – it is perhaps the most difficult of the three to come to terms with. There are two reasons for being interested in this case particularly. One reason is that, if you take the observations as they exist at the moment at their face value, it is the preferred model. According to General Relativity, the curvature of space is determined by the amount of matter present in the Universe and there doesn't seem to be enough to close the geometry of the Universe. Now, it may be that there is a lot of dark or hidden matter, which we do not yet know about. In this case, the Universe could be one of the other models but, if there is not a lot of extra matter, much more than we believe must be present within the optical images

28

Fig. 1.17. 'Circle Limit 4' by M. C. Escher (a representation
of Lobachevskian space).

of galaxies, then the Universe would have $k = -1$. The other reason is that it is the one I like the best! The properties of $k = -1$
geometries are particularly elegant.

What do the $k = -1$ universes look like? Their spatial sections
have what is known as hyperbolic or Lobachevski geometry. To
get a picture of a Lobachevski geometry it is best to look at one
of Escher's prints. He made a number of prints which he called
Circle Limits, and Circle Limit 4 is shown in Figure 1.17. This is
Escher's description of the Universe – you see it is full of angels
and devils! A point to note is that it looks as though the picture
gets very crowded towards the edge of the limit circle. This occurs
because this representation of hyperbolic space is drawn on an

ordinary plane sheet of paper, in other words, in Euclidean space. What you have to imagine is that all the devils are supposed to be actually exactly the same size and shape so that, if you happened to live in this Universe towards the edge of the diagram, they would look exactly the same to you as the ones in the middle of the diagram. This picture gives some impression of what is going on in Lobachevski geometry – as you walk from the centre out to the edge, you have to imagine that, because of the way the picture of the geometry has had to be distorted, the actual geometry there is exactly the same as it is in the middle, so that the geometry all about you remains the same no matter how you move.

This is perhaps the most surprising example of a well-defined geometry. But Euclidean geometry is, in its way, just as remarkable. Euclidean geometry provides a wonderful illustration of the relation between mathematics and physics. This geometry is a part of mathematics but the Greeks also thought of it as a description of the way the world is. Indeed, it turns out to be an extraordinarily accurate description of the way the world actually is – not utterly accurate because Einstein's theory tells us that space-time is slightly curved in various ways, but it is an extraordinarily accurate description of the world nevertheless. People used to worry about whether or not other geometries were possible. In particular, they worried about what is known as *Euclid's fifth postulate*. This can be reformulated as the statement that, if there is a straight line in a plane and there is a point outside that line, then there is a unique parallel to this line through that point. People used to think that maybe this could be proved from the other more obvious axioms of Euclidean geometry. It turns out that it is not possible, and from this the notion of non-Euclidean geometry arose.

In non-Euclidean geometries, the sum of the angles of a triangle do not add up to 180°. This is another example where you

(a)

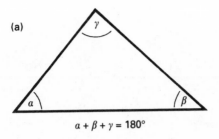

$$a + \beta + \gamma = 180°$$

(b)

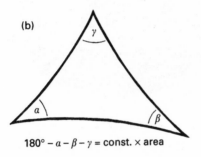

$$180° - a - \beta - \gamma = \text{const.} \times \text{area}$$

Fig. 1.18. (a) A triangle in Euclidean space. (b) A triangle in Lobachevskian space.

think that things must become more complicated because, in Euclidean geometry, the angles of a triangle do add up to 180° (Figure 1.18(a)). But then, in the non-Euclidean geometry, if you take the sum of the angles of a triangle away from 180°, you find that this difference is proportional to the area of the triangle. In Euclidean geometry, the area of a triangle is a complicated thing to write down in terms of angles and lengths. In non-Euclidean, Lobachevskian, geometry, there is this wonderfully simple formula, due to Lambert, which enables the area of the triangle to be found (Figure 1.18(b)). In fact, Lambert derived his formula

31

Fig. 1.19. 'Circle Limit 1' by M. C. Escher.

before non-Euclidean geometry was discovered and I have never quite understood that!

There is another very important point here, which concerns the real numbers. These are absolutely fundamental to Euclidean geometry. They were essentially introduced by Eudoxus in the fourth century BC and they are still with us. They are the numbers which describe all our physics. As we shall see later, complex numbers are needed too, but they are based upon real numbers.

Let us look at another of the Escher prints to see how the Lobachevski geometry works. Figure 1.19 is even nicer than Figure 1.17 for understanding this geometry because the 'straight lines' are more obvious. They are represented by arcs of circles which meet the boundary at right angles. So, if you were

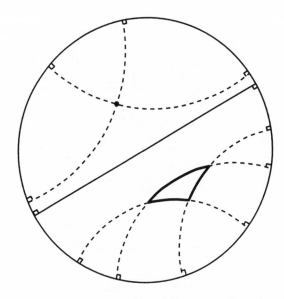

Fig. 1.20. Aspects of the geometry of Lobachevskian
(hyperbolic) space as illustrated by 'Circle Limit 1'.

a Lobachevskian person, and lived in this geometry, what you
would think of as a straight line would be one of these arcs. You
can see these clearly in Figure 1.19 – some of them are Euclidean
straight lines through the centre of the diagram but all the others
are curved arcs. Some of these 'straight lines' are shown in Figure
1.20. In that diagram, I have marked a point which does not lie on
the straight line (diameter) crossing the diagram. Lobachevskian
people can draw two (and more) separate lines parallel to the diameter through that point, as I have indicated. Thus, the parallel
postulate is violated in this geometry. Furthermore, you can draw
triangles and work out the sums of angles of the triangles in order to work out their areas. This may give you some taste for the
nature of hyperbolic geometry.

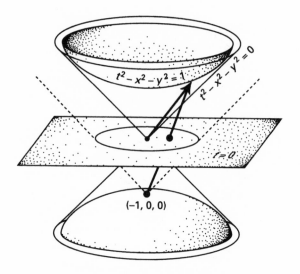

Fig. 1.21. Lobachevskian space embedded as a
hyperboloid branch in Minkowski space-time.
Stereographic projection takes it to the Poincaré disc,
whose boundary is the circle drawn on the plane $t = 0$.

Let me give another example. I said that I like hyperbolic,
Lobachevskian geometry the best. One of the reasons is that its
group of symmetries is exactly the same as the one that we have
already encountered, namely, the Lorentz group – the group of
Special Relativity, or the symmetry group of the light cones of relativity. To see that it is, I have drawn a light cone in Figure 1.21
but with some extra bits drawn on. I have had to suppress one of
the space dimensions in order to depict it in three-dimensional
space. The light cone is described by the usual equation shown
on the diagram

$$t^2 - x^2 - y^2 = 0.$$

The bowl-shaped surfaces shown above and below are located

at 'unit distance' from the origin in this Minkowskian geometry. ('Distance' in Minkowskian geometry is actually *time* – the proper time that is physically measured by moving clocks.) Thus, these surfaces represent the surface of a 'sphere' for the Minkowskian geometry. It turns out that the intrinsic geometry of the 'sphere' is actually Lobachevskian (hyperbolic) geometry. If you consider an ordinary sphere in Euclidean space, you can rotate it around, and the group of symmetries is that of rotating the sphere around. In the geometry of Figure 1.21, the group of symmetries is the group of symmetries associated with the surface shown in the diagram, in other words, with the Lorentz group of rotations. This symmetry group describes how space and time transform when a particular point in space-time is fixed – rotating the space-time about in different ways. We now see, with this representation, that the group of symmetries of Lobachevskian space is essentially just the same as the Lorentz group.

Figure 1.21 illustrates a Minkowskian version of the stereo-graphic projection shown in Figure 1.10(*c*). The equivalent of the south pole is now the point at $(-1, 0, 0)$ and we project points from the upper bowl-shaped surface to the flat surface at $t = 0$, which is the analogue of the equatorial plane in Figure 1.10(c). In this procedure, we project all the points on the upper surface to the plane at $t = 0$. The projected points all lie inside a disc in the plane at $t = 0$, and this disc is sometimes referred to as the Poincaré disc. This is precisely how Escher's circle limit diagrams come about – the entire hyperbolic (Lobachevskian) surface has been mapped onto the Poincaré disc. Furthermore, this mapping does all the things that the projection of Figure 1.10(c) does – it preserves angles and circles and it all comes out geometrically in a very nice way. Well, perhaps I am getting carried away here by my enthusiasm – I am afraid that is what mathematicians do when they get stuck into something!

The intriguing point is that, when you get carried away by something like the geometry of the above problem, the analysis and the results have an elegance which sustains them, while analyses which do not possess this mathematical elegance peter out. There is something particularly elegant about hyperbolic geometry. It would be awfully nice, at least to the likes of me, if the Universe were built that way too. Let me say that I have various other reasons for believing this. Many other people do not like these open, hyperbolic universes – they frequently prefer closed universes, such as those illustrated in Figure 1.16(b), which are nice and cosy. Well, actually, the closed universes are pretty big still. Alternatively, many people like flat world models (Figure 1.16(a)) because there is a certain type of theory of the early Universe, the *inflationary theory*, which suggests that the geometry of the Universe should be flat. I should say that I do not really believe these theories.

The three standard types of model of the Universe are known as the *Friedman models* and they are characterised by the fact that they are very, very symmetrical. They are initially expanding models but at any moment the Universe is perfectly uniform everywhere. This assumption is built into the structure of the Friedman models and it is known as the *cosmological principle*. Wherever you are, the Friedman universe looks the same in all directions. It turns out that our actual Universe is like this to a remarkable degree. If Einstein's equations are right, and I have shown that the theory agrees with observation to a quite extraordinary degree, then we are led to take the Friedman models seriously. All these models have this awkward feature, known as the Big Bang, where everything goes wrong, right at the beginning. The Universe is infinitely dense, infinitely hot and so on – something has gone badly wrong with the theory. Nonetheless, if you accept that this very hot, dense phase took place, you can

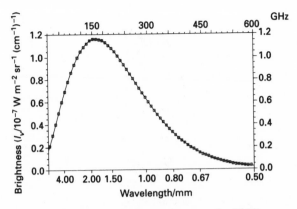

Fig. 1.22. The precise agreement between the COBE measurements of the spectrum of the Cosmic Microwave Background Radiation and the expected 'thermal' nature of the Big Bang's radiation (solid line).

make predictions about what the thermal content of the Universe should be today and one of these expectations is that there should be a uniform background of black-body radiation all about us at the present day. Precisely this type of radiation was discovered by Penzias and Wilson in 1965. The most recent observations of the spectrum of this radiation, which is known as the Cosmic Microwave Background Radiation, by the COBE satellite show that it has a black-body spectrum of quite extraordinary precision (Figure 1.22).

All cosmologists interpret the existence of this radiation as evidence that our Universe went through a hot, dense phase. This radiation is thus telling us something about the nature of the early Universe – it is not telling us everything, but something like the Big Bang did take place. In other words, the Universe must have been very like the models illustrated in Figure 1.16.

There is one other very important discovery made by the COBE satellite and that is that, although the Cosmic Microwave Back-

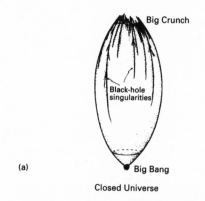

(a)

Closed Universe

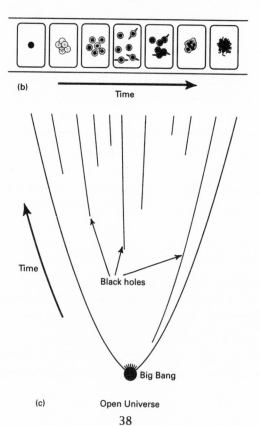

(b)

(c)   Open Universe

ground Radiation is remarkably uniform and its properties can all be accounted for very beautifully mathematically, the Universe is not quite perfectly uniform. There are tiny but real irregularities in the distribution of the radiation over the sky. In fact, we expect that these tiny irregularities must be present in the early Universe – we are here to observe the Universe and we are certainly not just a uniform smudge. The Universe is probably more like the pictures illustrated in Figure 1.23. To show how open-minded I am, I am using as examples both an open and a closed Universe.

In the closed Universe, the irregularities will develop to form real observable structures – stars, galaxies and the like – and, after a while, black holes will form, through the collapse of stars, through the accumulation of mass at the centres of galaxies and so on. These black holes all have singular centres, much like the Big Bang in reverse. However, it is not as simple as that. According to the picture we have developed, the initial Big Bang is a nice, symmetrical, uniform state but the end point of the closed model is a horrible mess – all the black holes finally coming together and producing an incredible jumble at the final Big Crunch (Figure 1.23(a)). The evolution of this closed model is illustrated schematically by the film strip shown in Figure 1.23(b). In the case of an open universe model, the black holes are still formed – there is still an initial singularity and singularities are formed at the centres of the black holes (Figure 1.23(c)).

Fig. 1.23. (a) The evolution of a closed world model with the formation of black holes, as objects of various types reach the end points of their evolution. It can be seen that there is expected to be a horrible mess at the Big Crunch. This sequence of events for (a) is also shown as a 'film-strip' in (b). (c) The evolution of an open model showing the formation of black holes at different times.

Time

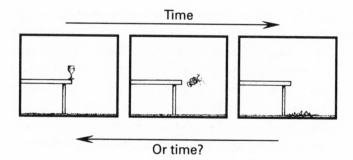

Or time?

Fig. 1.24. The laws of mechanics are time-reversible; yet the time-ordering of such a scene from the right frame to the left is something that is never experienced, whereas that from the left frame to the right would be commonplace.

I emphasise these features of the standard Friedman models to show that there is a great difference between what we seem to see in the initial state and what we expect to find in the remote future. This problem is connected with the fundamental law of physics known as the Second Law of Thermodynamics.

We can understand this law in simple everyday terms. Imagine a glass of wine perched on the edge of a table. It might fall off the table, smash to pieces and the wine spill all over the carpet (Figure 1.24). There is nothing in Newtonian physics which tells us that the reverse process cannot happen. However, that is never observed – you never see wine glasses reassembling themselves and the wine being sucked up out of the carpet and into the reassembled glass. So far as the detailed laws of physics are concerned, one direction of time is just as good as the other. To understand this difference, we need the Second Law of Thermodynamics which tells us that the entropy of the system increases with time. This quantity called entropy is lower when the

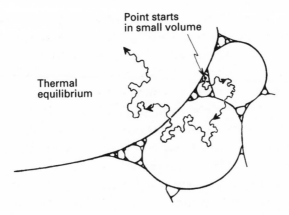

Point starts
in small volume

Thermal
equilibrium

Fig. 1.25. The Second Law of Thermodynamics in action:
as time evolves, the phase-space point enters
compartments of larger and larger volumes.
Consequently, the entropy continually increases.

glass is on the table as compared with when it is shattered on
the floor. According to the Second Law of Thermodynamics, the
entropy of the system has increased. Roughly speaking, entropy
is a measure of the disorder of a system. To express this concept more precisely, we have to introduce the concept of a *phase
space.*

A phase space is a space of an enormous number of dimensions and each point of this multi-dimensional space describes
the positions and momenta of all the particles which make up the
system under consideration. In Figure 1.25, we have selected a
particular point in this huge phase space which represents where
all the particles are located and how they are moving. As the system of particles evolves, the point moves to somewhere else in the
phase space and I have shown it wiggling about from one point
in phase space to another.

41

This wiggly line represents the ordinary evolution of the system of particles. There is no entropy there yet. To get entropy, we have to draw little bubbles around regions by lumping together those different states which you cannot tell apart. That may seem a bit obscure – what do you mean by 'cannot tell apart'? Surely that depends upon who are looking and how carefully they look? Well, it is one of the slightly tricky questions of theoretical physics to say exactly what you mean by entropy. Essentially, what is meant is that you have to group states together according to what is known as 'coarse-graining', that is, according to those things which you cannot tell apart. You take all those which, say, lie in this region of phase space here, lump them together, you look at the volume of that region of phase space, take the logarithm of the volume and multiply it by the constant known as Boltzmann's constant and that is the entropy. What the Second Law of Thermodynamics tells us is that the entropy increases. What it is telling you is actually something rather silly – all it is saying is that, if the system starts off in a little tiny box and it is allowed to envolve, it moves into bigger and bigger boxes. It is very likely that this happens because, if you look at the problem carefully, the bigger boxes are absolutely stupendously huger than the neighbouring little boxes. So, if you find yourself in one of the big boxes, there is virtually no chance at all of getting back into a smaller box. That is all there is to it. The system just wanders about in phase space getting into bigger and bigger boxes. That is what the Second Law is telling us. Or is it?

Actually, that is only half the explanation. It tells us that, if we know the state of the system now, we can tell the most likely state in the future. But it tells us the completely wrong answer if we try to use the same argument backwards. Suppose the glass is sitting on the edge of the table. We can ask 'What is the most likely way by which it could have got there?' If you use the argument

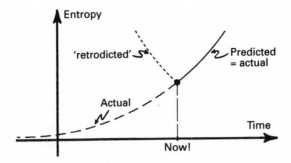

Fig. 1.26. If we use the argument depicted in Figure 1.25 in the reverse direction in time, we 'retrodict' that the entropy should also increase in the *past*, from its value now. This is in gross contradiction with observation.

we have just given backwards, you would conclude that the most likely thing is that it started as a great mess on the carpet and then picked itself up off the carpet and reassembled itself on the table. This is clearly not the correct explanation – the correct explanation is that someone put it there. And that person put it there for some reason, which was in turn due to some other reason and so on. The chain of reasoning goes back and back to lower and lower entropy states in the past. The correct physical curve is the 'actual' one illustrated in Figure 1.26 (not the 'retrodicted' one) – the entropy goes down and down and down in the past.

Why the entropy increases in the future is explained by moving into larger and larger boxes – why it goes down in the past is something completely different. There must have been something which pulled it down in the past. What pulled it down in the past? As we go into the past, the entropy gets smaller and smaller until eventually we end up at the Big Bang.

There must have been something very, very special about the Big Bang, but exactly what that was is a controversial issue. One

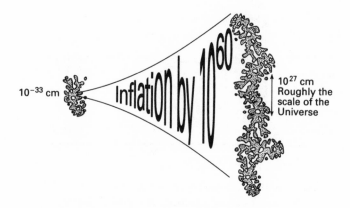

Fig. 1.27. Illustrating the problem of the inflation of
'generic' irregularities in the early Universe.

popular theory, which I said I did not believe but which a lot of
people are very keen on, is the idea of the inflationary universe.
The idea is that the Universe is so uniform on the large scale because of something which was supposed to have taken place in the
very earliest phases of the expansion of the Universe. It is supposed that an absolutely enormous expansion took place when
the Universe was only about 10^{-36} seconds old and the idea is
that, no matter what the Universe looked like in these very early
stages, if you expand it up by a huge factor of about 10^{60}, then it
will look flat. In fact, this is one reason why these people like the
flat Universe.

But, as it stands, the argument does not do what it is supposed
to do – what you would expect in this initial state, if it were ran-
domly chosen, would be a horrendous mess and, if you expand
that mess by this huge factor, it still remains a complete mess. In
fact, it looks worse and worse the more it expands (Figure 1.27).

So the argument by itself does not explain why the Universe is
so uniform. We need a theory which tells you what the Big Bang

was really like. We do not know what that theory is but we know that it has to involve a combination of large-scale and small-scale physics. It has to involve quantum physics as well as classical physics. Furthermore, I would claim that the theory must also have as one of its implications that the Big Bang was as uniform as we observe it to be. Maybe such a theory will end up producing a hyperbolic, Lobachevskian universe, like the picture I prefer, but I shall not insist upon that.

Let us return to the pictures of the closed and open universes again (Figure 1.28). In addition, I have included a picture of the formation of a black hole, which will be well known to the experts. Matter collapsing into a black hole produces a singularity and that is what the dark lines on the space-time diagrams of the Univese represent. I want to introduce a hypothesis which I call the *Weyl curvature hypothesis*. This is not an implication of any known theory. As I have said, we do not know what the theory is, because we do not know how to combine the physics of the very large and the very small. When we do discover that theory, it should have as one of its consequences this feature which I have called the Weyl curvature hypothesis. Remember that the Weyl curvature is that bit of the Riemann tensor which causes distortions and tidal effects. For some reason we do not yet understand, in the neighbourhood of the Big Bang, the appropriate combination of theories must result in the Weyl tensor being essentially zero, or rather being constrained to be very small indeed.

That would give us a Universe like that shown in Figure 1.28(a) or (c) and not like that in Figure 1.29. The Weyl curvature hypothesis is time-asymmetrical and it applies only to the past type singularities and not to the future singularities. If the same flexibility of allowing the Weyl tensor to be 'general' that I have applied in the future also applied to the past of the Universe, in the closed model, you would end up with a dreadful looking Universe with

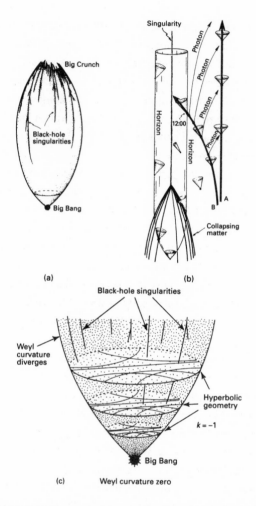

(a)

(b)

(c) Weyl curvature zero

Fig. 1.28. (a) The entire history of a closed universe which starts from a uniform low-entropy Big Bang with **Weyl** = 0 and ends with a high-entropy Big Crunch – representing the congealing of many black holes – with **Weyl** → ∞. (b) A space-time diagram depicting collapse to an individual black hole. (c) The history of an open universe, again starting from a uniform low-entropy Big Bang with **Weyl** = 0.

46

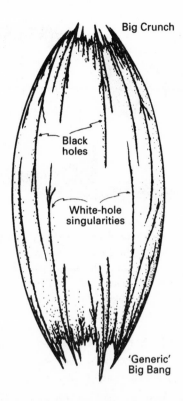

Fig. 1.29. If the constraint **Weyl** = 0 is removed, then we have a high-entropy Big Bang also, with **Weyl** → ∞ there. Such a universe would be riddled with white holes, and there would be no Second Law of Thermodynamics, in gross contradiction with experience.

as much mess in the past as in the future (Figure 1.29). This looks nothing like the Universe we live in.

What is the probability that, purely by *chance*, the Universe had an initial singularity looking even remotely as it does? The probability is less than one part in $10^{10^{123}}$. Where does this estimate

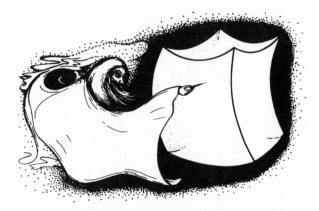

Fig. 1.30. In order to produce a universe resembling the one in which we live, the Creator would have to aim for an absurdly tiny volume of the phase space of possible universes – at most about $1/10^{10^{123}}$ of the entire volume. (The pin, and the spot aimed for, are not drawn to scale!)

come from? It is derived from a formula by Jacob Beckenstein and Stephen Hawking concerning black-hole entropy and, if you apply it in this particular context, you obtain this enormous answer. It depends how big the Universe is and, if you adopt my own favourite Universe, the number is, in fact, infinite.

What does this say about the precision that must be involved in setting up the Big Bang? It is really very, very extraordinary. I have illustrated the probability in a cartoon of the Creator, finding a very tiny point in that phase space which represents the initial conditions from which our Universe must have evolved if it is to resemble remotely the one we live in (Figure 1.30). To find it, the Creator has to locate that point in phase space to an accuracy of one part in $10^{10^{123}}$. If I were to put one zero on each elementary particle in the Universe, I still could not write the number down in full. It is a stupendous number.

I have been talking about precision – how mathematics and physics fit together with extraordinary accuracy. I have also talked about the Second Law of Thermodynamics, which is often thought of as a rather floppy law – it concerns randomness and chance – and yet there is something very precise hiding underneath this law. As applied to the Universe, it has to do with the precision with which the initial state was set up. This precision must be something to do with the union of quantum theory and general relativity, a theory we do not have. In the next chapter, however, I shall tell you something about the sorts of things which should be involved in such a theory.

CHAPTER 2

The Mysteries of Quantum Physics

In the first chapter, I made the case that the structure of the physical world is dependent, very precisely, upon mathematics, as illustrated symbolically in Figure 1.3. It is remarkable how extraordinarily precise mathematics is in describing the most fundamental aspects of physics. In a famous lecture, Eugene Wigner (1960) referred to this as:

> The unreasonable effectiveness of mathematics in the physical sciences.

The list of successes is very impressive:

Euclidean geometry is accurate to smaller than the width of a hydrogen atom over a metre's range. As discussed in the first lecture, it is not precisely accurate because of the effects of General Relativity but nonetheless, for most practical purposes, Euclidean geometry is very precise indeed.

Newtonian mechanics is known to be accurate to about one part in 10^7 but it is not precisely accurate – again, we need relativity to obtain more accurate results.

Maxwell's electrodynamics holds good over an enormous range of scales from the sizes of particles, when taken in conjunction with quantum mechanics, out to the sizes of distant galaxies, corresponding to a range of scales of 10^{35} or more.

Einstein's Relativity, as discussed in the first chapter, can be said to be accurate to about one part in 10^{14}, roughly twice as many figures as Newtonian mechanics, where Einstein's theory is taken to include Newtonian mechanics.

Quantum mechanics is the subject of this chapter and is also an extraordinarily accurate theory. In quantum field theory, which is the combination of quantum mechanics with Maxwell's electrodynamics and Einstein's Special Theory of relativity, there are effects which can be computed to be accurate to about one part in 10^{11}. Specifically, in a set of units known as 'Dirac units', the magnetic moment of the electron is predicted to be 1.001159652(46), compared with the experimentally determined value of 1.0011596521(93).

There is an important point concerning these theories – not only is the mathematics quite extraordinarily effective and accurate in its description of our physical world, but it is also extraordinarily fruitful as mathematics itself. Very often one finds that some of the most fruitful concepts in mathematics have been based upon concepts which have come out of physical theories. Here are some examples of the types of mathematics which have been stimulated by the requirements of physical theories:

- real numbers;
- Euclidean geometry;
- calculus and differential equations;
- symplectic geometry;
- differential forms and partial differential equations;
- Riemannian geometry and Minkowski geometry;
- complex numbers;

- Hilbert space;
- functional integrals;
  ... and so on.

One of the most striking examples was the discovery of calculus, which was developed by Newton and others in order to provide the mathematical foundations of what we now call Newtonian mechanics. When these various types of mathematics were subsequently applied to the solution of purely mathematical problems, they also turned out to be extremely fruitful as mathematics *per se*.

In Chapter 1, we examined the scales of objects, ranging from the Planck length and the Planck time, the fundamental units of length and time, through the smallest sizes encountered in particle physics, which are about 10^{20} times greater than the Planck scale, through the human length-scale and time-scale, showing that we are extremely stable structures in the Universe, right up to the age and radius of our physical Universe. I mentioned the rather disturbing fact that, in our description of fundamental physics, we use two quite different ways of describing the world, depending upon whether we are talking about the large-scale or the small-scale end of things. Figure 2.1 (which is a reproduction of Figure 1.5) illustrates that we use quantum mechanics to describe the small quantum level of activity and classical physics to describe phenomena on the large scale. I have denoted these levels of activity as U for the quantum level, standing for Unitary, and C for the Classical level. I discussed large-scale physics in Chapter 1 and emphasised that we seem to have quite different laws on the large and the small scale.

I believe that the normal view of physicists is that, if we really understood quantum physics properly, we could deduce classical physics from it. I want to argue differently. In practice, one does

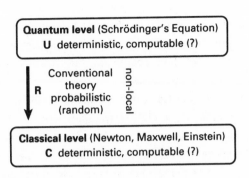

Fig. 2.1.

not do that – one uses *either* the classical level *or* the quantum level. This is disturbingly like how the ancient Greeks looked at the world. To them, there was one set of laws which applied on the Earth and a different set of laws which applied in the heavens. It was the strength of the Galilean–Newtonian viewpoint that one could bring these two sets of laws together and see that they could be understood in terms of the same physics. We now seem to be back with a kind of Greek situation, with one set of laws applying at the quantum level and another set at the classical level.

There is one possible misunderstanding which I should clear up concerning Figure 2.1. I have put the names of Newton, Maxwell and Einstein in the box labelled 'Classical level', along with the word 'deterministic'. I do not mean that they believed, for example, that the way the Universe behaves is deterministic. It is quite reasonable to suppose that Newton and Maxwell did not hold that view, although Einstein apparently did. The remarks 'deterministic, computable(?)' refer to their theories only and not to what the scientists believed about the actual world. In the box labelled 'Quantum level', I have included the words 'Schrödinger's Equation' and, I am sure, he did not believe that all physics is described by the equation named after him. I shall come back to this point later. In other words, the people and the theories named after them are quite separate things.

Well, are there really these two distinct levels illustrated in Figure 2.1? We certainly might ask the question: 'Is the Universe precisely governed by quantum mechanical laws alone? Can we explain the entire Universe in terms of quantum mechanics?' To address this question, I shall have to say something about quantum mechanics. Let me give first a brief list of some of the things that quantum mechanics can explain.

- *The stability of atoms* Before the discovery of quantum mechanics, it was not understood why the electrons in atoms did not spiral into their nuclei, as they should according to an entirely classical description. Stable classical atoms should not exist.

- *Spectral lines* The existence of quantised energy levels in atoms and the transitions between them give rise to the emission lines we observe with precisely defined wavelengths.

- *Chemical forces* The forces which hold molecules together are entirely quantum mechanical in nature.

- *Black-body radiation* The spectrum of black-body radiation can only be understood if the radiation itself is quantised.

- *The reliability of inheritance* This depends upon quantum mechanics at the molecular level of DNA.

- *Lasers* The operation of lasers depends upon the existence of stimulated quantum transitions between quantum mechanical states of molecules and on the quantum (Bose–Einstein) nature of light.

- *Superconductors and superfluids* These are phenomena which occur at very low temperatures and are associated with long-range quantum correlations between electrons (and other particles) in various substances.

- ...etc. ...etc.

In other words, quantum mechanics is omnipresent even in everyday life and is at the heart of many areas of high technology, including electronic computers. *Quantum Field Theory*, the combination of quantum mechanics with Einstein's Special Theory of relativity, is also essential for understanding particle physics. As mentioned above, Quantum Field Theory is known to be accurate to about one part in 10^{11}. This list indicates just how wonderful and powerful quantum mechanics is.

Let me say a little bit about what quantum mechanics is. The archetypal experiment is shown in Figure 2.2. According to quantum mechanics, light consists of particles called *photons*, and the figure shows a photon source which we assume emits photons one at a time. There are two slits t and b and a screen behind them. The photons arrive at the screen as individual events, where they are detected separately, just as if they were ordinary particles. The curious quantum behaviour arises in the following way. If only slit t were open and the other closed, there would be many places on the screen which the photon could reach. If I now close the slit t and open slit b, I may again find that the photon could reach the same spot on the screen. But, if I open both slits, and if I have chosen my point on the screen carefully, I may now find that the photon cannot reach that spot, even though it could have done so if either slit alone were open. Somehow, the two possible things which the photon *might* do cancel each other out. This type of behaviour does not take place in classical physics. Either one thing happens or another thing happens – you don't get two possible things which might happen, somehow conspiring to cancel each other out.

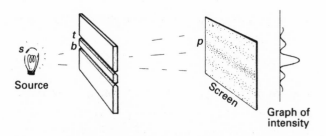

Fig. 2.2. The two-slit experiment, with individual photons of monochromatic light.

The way we understand the outcome of this experiment in quantum theory is to say that when the photon is *en route* from the source to the screen, the state of the photon is not that of having gone through one slit or the other, but is some mysterious combination of the two, weighted by *complex numbers*. That is, we can write the state of the photon as

$$\mathbf{w} \times (\text{alternative A}) + \mathbf{z} \times (\text{alternative B})$$

where **w** and **z** are complex numbers. (Here 'alternative A' might stand for the route stp taken by the photon, in Figure 2.2, with 'alternative B' standing for the route sbp.)

Now, it is important that the numbers multiplying the two alternatives are complex numbers – that is the reason why the cancellations occur. You might think that you could work out the behaviour of the photon in terms of the probability that it did one thing or another, and then **w** and **z** would be real-number probability weightings. But this interpretation is not correct because **w** and **z** are complex. This is the important thing about quantum mechanics. You cannot explain the wavelike nature of quantum particles in terms of 'probability waves' of alternatives. They are *complex waves* of alternatives! Now, complex numbers are things which involve the square root of minus one, $i = \sqrt{(-1)}$, as well as the ordinary real numbers. They can be represented on a two-dimensional plot with the purely real numbers running along the x-axis, the real axis, and the purely imaginary numbers running up the y-axis, the imaginary axis, as illustrated in Figure 2.3(a). In general, a complex number is some combination of purely real and purely imaginary numbers, such as $2 + 3\sqrt{(-1)} = 2 + 3i$, and can be represented by a point in the plot of Figure 2.3(a), often known as an Argand diagram (or Wessel plane or Gauss plane).

Each complex number can be represented as a point in Figure 2.3(a) and there are various rules about how you add them, multiply them and so on. For example, to add them, you just use the parallelogram rule, which amounts to adding together the real and imaginary parts separately, as illustrated in Figure 2.3(b). You can also multiply them together, using the similar-triangle rule, as illustrated in Figure 2.3(c). When you become familiar with diagrams like those in Figure 2.3, the complex numbers become much more concrete things, rather than abstract objects. The fact that these numbers are built into the foundations of quantum theory often makes people feel that the theory is a rather abstract and unknowable kind of thing, but once you get used to complex numbers, particularly after playing around with them on the Argand diagram, they become very concrete objects and you don't worry so much about them.

There is, however, more to quantum theory than simply the superposition of states weighted by complex numbers. So far, we have remained at the quantum level, where the rules that I am calling U apply. At that level, the state of the system is given by a complex-number weighted superposition of all possible alternatives. The time-evolution of the quantum state is called *unitary evolution* (or Schrödinger evolution) – which is what U actu-

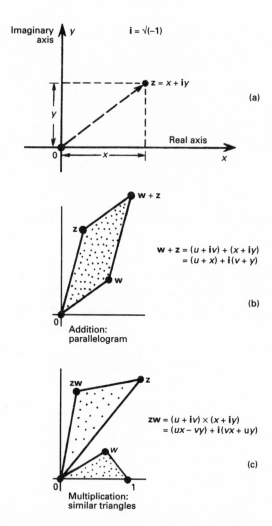

Fig. 2.3. (a) The representation of a complex number in the (Wessel–Argand–Gauss) complex plane. (b) The geometrical description of the addition of complex numbers. (c) The geometrical description of the multiplication of complex numbers.

ally stands for. An important property of U is that it is *linear*. This means that a superposition of two states always evolves in the same way as each of the two states would individually, but superposed together with complex-number weightings that remain *constant in time*. This linearity is a fundamental feature of Schrödinger's Equation. At the quantum level, these complex-number weighted superpositions are always maintained.

When you magnify something to the classical level, however, you then *change the rules*. By magnifying to the clasical level, I mean going from the top level U to the bottom level C of Figure 2.1 – physically this is what happens, for example, when you observe a spot on the screen. A small-scale quantum event triggers something larger that can actually be seen at the classical level. What you do in standard quantum theory is to wheel out of the cupboard something which people do not like to mention too much. It is what is called the *collapse of the wavefunction* or the *reduction of the state vector* – I am using the letter **R** for this process. You do something completely different from unitary evolution. In a superposition of two alternatives, you look at the two complex numbers and you take the squares of their moduli – that means taking the squares of the distances from the origin of the two points in the Argand plane – and these two squared moduli become the ratios of the probabilities of the two alternatives. But this only happens when you 'make a measurement', or 'make an observation'. One can think of this as the process of magnifying phenomena from the U to the C levels of Figure 2.1. With this process, you change the rules – you no longer maintain these linear superpositions. Suddenly, the ratios of these squared moduli become probabilities. It is only in going from the U to the C level that you introduce non-determinism. This non-determinism comes in with **R**. Everything at the U level is deterministic – quantum mechanics only becomes non-deterministic when you do this thing which is called 'making a measurement'.

So, this is the scheme one uses in standard quantum mechanics. It is a very odd type of scheme for a fundamental theory. Maybe if it were only an approximation to some other more fundamental theory, it might make more sense, but this hybrid procedure is itself regarded by all the professionals as a fundamental theory!

Let me say a little bit more about these complex numbers. At first, they seem very abstract things which sit around until you square their moduli and then they become probabilities. In fact, they often have a strongly geometrical character. I want to give you an example in which their meaning can be appreciated more clearly. Before I do that, let me say a little more about quantum mechanics. I shall use these funny-looking brackets, known as Dirac brackets. They are simply a shorthand for describing the state of the system – when I write $|A\rangle$, I mean that the system is in quantum state A. What sits inside the bracket is some description of the quantum state. Very often, the overall quantum mechanical state of the system is written as ψ, which is some superposition of other states and we might write this as

$$|\psi\rangle = \mathbf{w}|A\rangle + \mathbf{z}|B\rangle$$

for the case of the two-slit experiment.

Now, in quantum mechanics, we are not so interested in the sizes of the numbers themselves as we are in their ratio. There is a rule in quantum mechanics that you are allowed to multiply the state by some complex number and that does not change the physical situation (so long as that complex number is not zero). In other words, it is only the ratio of these complex numbers which has direct physical meaning. When R comes in, we look at the probabilities, and then it is the ratios of the squared moduli which are needed, but if we stay at the quantum level, we can also hope

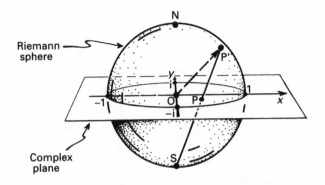

Fig. 2.4. The Riemann sphere. The point P, representing
$\mathbf{u} = \mathbf{z}/\mathbf{w}$ on the complex plane, is projected from the
south pole S to a point P' on the sphere. The direction
OP', from the sphere's centre O, is the direction of the
spin axis for the superposed state of two spin-$\frac{1}{2}$ particles.

to interpret the ratios of these complex numbers themselves, even
before their moduli are taken. The *Riemann sphere* is a way of
representing complex numbers on a sphere (Figure 1.10(c)). More
correctly, we are not just dealing with complex numbers but with
ratios of complex numbers. We have to be careful with ratios
because the thing in the denominator might turn out to be zero,
in which case that ratio comes out as infinity – we have to deal
with that case as well. We can place all the complex numbers,
together with infinity, on a sphere by this very neat projection in
which the Argand plane is now the equatorial plane, intersecting
the sphere in the unit circle, which is the sphere's equator (Figure
2.4). Evidently, we can project each point on the equatorial plane
onto the Riemann sphere, projecting from its south pole. As can
be seen from the diagram, the south pole of the Riemann sphere
would correspond, in this projection, to *'the point at infinity'* in
the Argand plane.

If a quantum system has two alternative states, the different states which can be made by combining these two are represented by a sphere – an abstract sphere at this stage – but there are circumstances in which you can actually see it. I am very attached to the following example. If we have a spin-$\frac{1}{2}$ particle, such as an electron, a proton, or a neutron, then the various combinations of their spin states can be realised geometrically. Spin-$\frac{1}{2}$ particles can have two spin states, one with the rotation vector pointing upwards (the up-state) and the other with the rotation vector pointing downwards (the down-state). The superposition of the two states can be represented symbolically by the equation

$$\left| \varnothing \right\rangle = \mathbf{w} \left| \updownarrow \right\rangle + \mathbf{z} \left| \updownarrow \right\rangle$$

The different combinations of these spin states give us rotation about some other axis and, if you want to know where that axis is, you take the ratio of the complex numbers $\mathbf{w}$ and $\mathbf{z}$, which gives you another complex number $\mathbf{u} = \mathbf{z}/\mathbf{w}$. You place this new number $\mathbf{u}$ on the Riemann sphere and the direction of that complex number from the centre is the direction of the spin axis. So you see, the complex numbers of quantum mechanics are not as abstract as they might seem at first. They have a quite concrete meaning – sometimes this meaning is a little bit hard to dig out, but in the case of the spin-$\frac{1}{2}$ particle, the meaning is manifest.

This analysis of spin-$\frac{1}{2}$ particles tells us something else. There is nothing special about spin-up and spin-down. I might have chosen any other axis I liked, say, left or right, or forward and backwards – it does not make any difference. This illustrates that there is nothing special about the two states you start with (except that the chosen two spin states should be the opposite of each other). According to the rules of quantum mechanics, any other spin state is on just as good a footing as each of the two you start with. This is clearly illustrated in this example.

Quantum mechanics is a beautiful, clear-cut subject. However, it also has many mysteries. It is certainly a mysterious subject and, in many different ways, a puzzling or paradoxical subject. I want to stress that there are mysteries of *two different kinds*. I call these the *Z* and *X* mysteries.

The *Z*-mysteries are the *puZZle* mysteries – they are things which are certainly there in the physical world, that is, there are good experiments which tell us that quantum mechanics does behave in these mysterious ways. Maybe some of these effects have not been fully tested but one has very little doubt that the quantum mechanics is right. These mysteries include phenomena such as the *wave–particle duality* I referred to earlier, *null measurements*, which I shall talk about in a moment, *spin*, which I only just talked about, and *non-local effects*, which I shall also talk about shortly. These things are genuinely puzzling phenomena but few people argue about their reality – they are certainly part of nature.

There are other problems, however, which I refer to as *X*-mysteries. These are the *paradoX* mysteries. These, to my way of thinking, are indications that the theory is incomplete, wrong or something else – it needs some further attention. The essential *X*-mystery concerns the *measurement problem*, which I have discussed above – namely, the fact that the rules change from U to R when we pass out of the quantum and enter the classical level. Could we understand why this R procedure arises, as perhaps an approximation, or illusion, if we better understood how large and complicated quantum systems behave? The most famous of the *X*-paradoxes concerns *Schrödinger's cat*. In this experiment – I stress, a thought experiment, since Schrödinger was a very humane man – the cat is in a state of being both dead and alive at the same time. You don't actually see cats like this. I shall say more about this problem in a moment.

My view is that we must learn to snooze happily with the Z-mysteries but the X-mysteries should be crossed off when we have a better theory. I stress that this is very much my own view of the X-mysteries. Many others view the (apparent?) paradoxes of quantum theory in a different light – or, I should say, in *many different* lights!

Let me say something about the Z-mysteries before I come to the more serious problems of the X-mysteries. I shall discuss two of the most striking of the Z-mysteries. One of these is the problem of *quantum non-locality* or, as some people prefer, *quantum entanglement*. It is a very extraordinary thing. The idea originally came from Einstein and his colleagues, Podolsky and Rosen, and is known as the EPR experiment. The version which is probably easiest to understand is that given by David Bohm. One has a particle of spin 0 which splits into two spin-$\frac{1}{2}$ particles, say an electron and positron going off in opposite directions. We then measure the spins of the particles that go off to widely separated points A and B. There is a very famous theorem due to John Bell which tells us that there is a conflict between the expectations of quantum mechanics concerning the joint probabilities of the results of measurements at the points A and B and any 'local realistic' model. By 'local realistic' model, I mean any model in which the electron is a thing at A and the positron is another thing at B, and those two things are separate from each other – they are not connected in any way. Then, that hypothesis gives results for the joint probabilities of measurement that might be performed at A and B which are in conflict with quantum mechanics. John Bell made this very clear. It is a very important result, and subsequent experiments, such as that performed by Alain Aspect in Paris, have confirmed this prediction of quantum mechanics. The

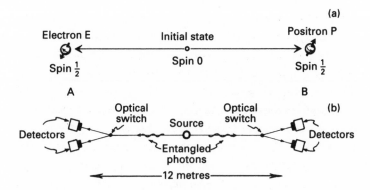

Fig. 2.5. (a) A spin-0 particle decays into two spin-$\frac{1}{2}$ particles, an electron E and a positron P. Measurement of the spin of one of the spin-$\frac{1}{2}$ particles apparently *instantaneously* fixes the spin state of the other. (b) The EPR experiment of Alain Aspect and colleagues. Photon pairs are emitted at the source in an entangled state. The decision as to which direction to measure each photon's polarization is not made until the photons are in full flight – too late for a message to reach the opposite photon, telling it of the direction of measurement.

experiment is illustrated in Figure 2.5 and concerns the polarisation states of pairs of photons emitted in opposite directions from a central source.

The decision as to which directions of polarisation of the photons were to be measured was not taken until the photons were in full flight from the source to the detectors at A and B. The results of these measurements clearly showed that the joint probabilities for the polarisation states of the photons detected at A and B agreed with the predictions of quantum mechanics, as most people, including Bell himself, would have believed, but in violation of the natural assumption that these two photons are separate independent objects. The Aspect experiment established quantum entanglement effects over a distance of about 12 metres. I am told that there are now some experiments to do with quantum cryptography in which similar effects take place over distances of the order of kilometres.

I should emphasise that, in these *non-local* effects, events occur at separated points at A and at B, but they are connected in mysterious ways. The way in which they are connected – or *entangled* – is a very subtle thing. They are entangled in such a way that there is no way of using that entanglement to send a signal from A to B – this is very important for the consistency of quantum theory with relativity. Otherwise, it would have been possible to use quantum entanglement to send messages faster than light. Quantum entanglement is a very strange type of thing. It is somewhere between objects being separate and being in communication with each other – it is a purely quantum mechanical phenomenon and there is no analogue of this in classical physics.

A second example of a Z-mystery concerns *null measurements*, and it is well illustrated by the *Elitzur–Vaidman bomb-testing problem.* Imagine you belong to a group of terrorists and you have come across a large collection of bombs. Each bomb has an ultra-sensitive detonator on its nose, so sensitive that a single visible light photon reflected off a little mirror at the end of its nose imparts a sufficient impulse to it to set the bomb off in a violent explosion. There is, however, a rather large proportion of duds among the whole collection of bombs. These are duds in a particular way. The trouble is that the delicate plunger, to which the mirror is attached, got stuck during manufacture and so, when a single photon hits the mirror of a dud bomb, it does not move the plunger and the bomb fails to go off (Figure 2.6(a)). The key point is that the mirror on the nose of the dud bomb now acts just as an ordinary fixed mirror, rather than a moveable one that acts as part of the detonation mechanism. So, here is the problem – find a guaranteed good bomb, given a large collection which includes a number of dud bombs. In classical physics, there is simply no way you can do that. The only way to test if it is a good bomb would be to wiggle the detonator and then the bomb explodes.

It is quite extraordinary that quantum mechanics enables you to test whether something *might* have happened but didn't happen. It tests what philosophers call *counterfactuals*. It is remarkable that quantum mechanics allows real effects to result from counterfactuals!

Let me show you how you can solve the problem. Figure 2.6(b) shows the original version of the solution provided by Elitzur and Vaidman in 1993. Suppose we have a dud bomb. It has a mirror which is jammed – it is just a fixed mirror – and so there is no significant wiggling of the mirror, and no explosion, when a photon bounces off it. We set up the arrangement shown in Figure 2.6(b). A photon is emitted which first encounters a half-silvered mirror. This is a mirror which transmits half the light incident upon it and reflects the other half. You might think that this means that half of the photons which encounter the mirror are transmitted through it and half bounce off. However, this is not at all what happens at the quantum level of single photons. In fact, each single photon, emitted individually from the source, would be put into a state of quantum superposition of both alternative routes for the photon: transmitted and reflected. The bomb's mirror is to lie in the path of the transmitted photon beam angled at 45°. The part of the photon beam that is reflected from the half-silvered mirror encounters another fully silvered mirror, also angled at 45°, and both beams then come togther at a final half-silvered mirror, as indicated in Figure 2.6(b). There are detectors at two places, A and B.

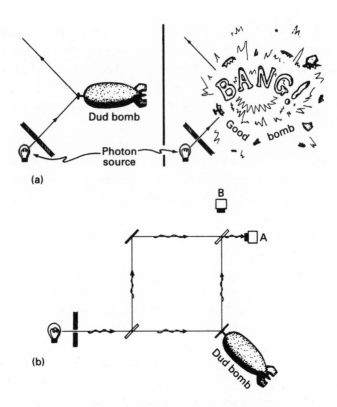

Fig. 2.6. (a) The Elitzur–Vaidman bomb-testing problem.
The bomb's ultra-sensitive detonator will respond to the
impulse of a single visible-light photon – assuming that
the bomb is not a dud because its detonator is jammed.
The problem is to find a guaranteed good bomb, given a
large supply of questionable ones. (b) The arrangement
for finding good bombs in the presence of duds. For
a good bomb, the lower right-hand mirror acts as a
measuring device. When it measures that a photon has
gone the other way, this allows the detector at B to receive
the photon – which cannot happen for a dud.

Let us consider what happens to a single photon, emitted by the source, the bomb being a dud. When it encounters the first half-silvered mirror, the state of the photon is divided into two separate states, one of which corresponds to the photon passing through the half-silvered mirror and heading towards the dud bomb and the other corresponding to the photon being reflected towards the fixed mirror. (This superposition of alternative photon routes is exactly the same as that which happens in the two-slit experiment illustrated in Figure 2.2. It is also essentially the same phenomenon as happens when we add spins together.) We suppose that the path lengths from the first to the second half-silvered mirror are exactly the same. To see what the photon state is when it reaches the detectors, we have to compare the two routes that the photon can take to reach either of the detectors, these two routes occurring in quantum superposition. We find the routes cancel out at B, whereas they add up at A. Thus, there can only be a signal to activate the detector A and never detector B. It is just like the interference pattern shown in Figure 2.2 – there are some positions at which there is never any intensity, because the two bits of the quantum state cancel at that point. Thus, on reflection from a dud bomb, the detector A is always activated, and never B.

Suppose we now have a good bomb. Then, the mirror on its nose is no longer a fixed mirror but its potentiality for wiggling turns the bomb into a *measuring device*. The bomb measures one or other of the two alternatives for the photon at the mirror – it can be in a state of a photon having arrived or not having arrived. Suppose the photon passes through the first half-silvered mirror and the mirror on the nose of the bomb measures that it has indeed come its way. Then, 'Boom!!!', the bomb goes off. We have lost it. So, we wheel in a new bomb and try again. Perhaps this time, the bomb measures that the photon does not arrive –

it does not explode, so the photon is measured to have gone the other way. (This is a null measurement.) Now, when the photon reaches the second half-silvered mirror, it is equally transmitted and reflected and so it is now possible for B to be activated. Thus, with a good bomb, every now and then, a photon is detected by B, indicating that the photon was measured by the bomb to have gone the other way. The key point is that, when the bomb is a good bomb, it acts as a measuring device, and this interferes with the exact cancellation that is necessary for preventing the photon from being detected by B, even though the photon does not interact with the bomb – a *null measurement*. If the photon did not come one way, then it had to have gone the other way! If B detects the photon, we know that the bomb acted as a measuring device and so it was a good bomb. Moreover, with a good bomb, every now and then, the detector B would measure the arrival of the photon and the bomb does not explode. This can *only* happen if it is a good bomb. You know that it is a good bomb because it has measured that the photon has actually gone the other way.

It really is extraordinary. In 1994, Zeilinger visited Oxford and told me that he had actually done the bomb-testing experiment. Actually, he and his collegues had not done it with bombs but with something similar, in principle – I should emphasise that Zeilinger is most certainly not a terrorist. He then told me that he and his colleagues Kwiat, Weinfurter and Kasevich had an im-proved solution in which they can actually do the same type of experiment without wasting any bombs at all. I shall not go into how that is done since it is a much more sophisticated arrange-ment. Actually, there is a vanishingly small amount of wastage, but, essentially with no wastage, you can find a guaranteed good bomb.

Let me leave you with these thoughts. These examples illustrate some aspects of the extraordinary nature of quantum mechanics

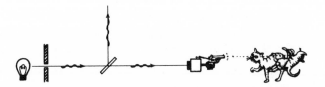

Fig. 2.7. *Schrödinger's cat*. The quantum state involves
a linear superposition of a reflected and transmitted
photon. The transmitted component triggers a device that
kills a cat, and so according to U-evolution the cat exists
in a superposition of life and death.

and its Z-mysteries. I think part of the trouble is that some people get hyponotised by these things – they say, 'Goodness me, quantum mechanics is that amazing' and indeed they are correct. It has to be amazing enough to include all these Z-mysteries as real phenomena. But then they think that they have to accept the X-mysteries too, and I believe that is wrong!

Let us turn to Schrödinger's cat. The version of the thought experiment shown in Figure 2.7 is not quite Schrödinger's original version but it will be more appropriate for our purposes. We again have a source of photons and a half-silvered mirror which splits the impinging photon's quantum state into a superposition of two different states, one reflected and the other passing through the mirror. There is a photon detection device in the path of the transmitted photon which registers the arrival of a photon by firing a gun which kills the cat. The cat may be thought of as the end point of a measurement; we change from the quantum level to the world of ponderable objects when the cat is found to be either dead or alive. But the problem is that if you take the quantum level as being true right the way up to the level of cats and so on, then you have to believe that the actual state of the cat is a superposition of both being dead and alive. The point is that the photon is in a superposition of states going one way or another, the detector is in a superposition of states of being off and on, and the cat in a superposition of states of being alive and dead. This problem has been known for a long time. What do various people say about it? There are probably more different attitudes to quantum mechanics than there are quantum physicists. This is not inconsistent because certain quantum physicists hold different views at the same time.

I want to illustrate a broad categorisation of viewpoints with a wonderful dinner-table remark made by Bob Wald. He said,

> If you really *believe* in quantum mechanics, then you
> cannot take it *seriously*.

It seems to me that this is a very true and profound remark about quantum mechanics and people's attitudes to it. I have divided quantum physicists into various categories in Figure 2.8. In particular, I have divided them into those who *believe* and those who are *serious*. What do I mean by serious? The serious people take the state vector $|\psi\rangle$ to describe the real world – the state vector *is* reality. Those who 'really' believe in quantum mechanics do not believe that this is the correct attitude to quantum mechanics. I have placed the names of various people on the diagram. So far as I can make out, Niels Bohr and followers of the Copenhagen viewpoint are believers. Bohr certainly believed in quantum mechanics but he did not take the state vector seriously as a description of the world. Somehow, $|\psi\rangle$ was all in the mind – it was our way of describing the world, it was not the world itself. And this also leads to what John Bell called FAPP, standing for 'For All Practical Purposes'. John Bell liked the term, I think because it has a slightly derogatory sound to it. It is based on the 'decoherence viewpoint' about which I shall have a little bit to say later. You

'If you really believe in quantum mechanics, then you cannot take it seriously.' (Bob Wald)

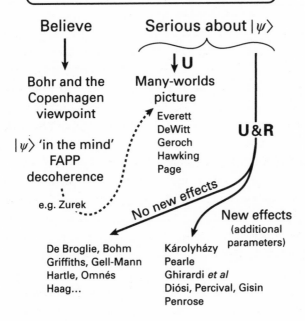

Fig. 2.8.

often find that, when you question thoroughly some of the most ardent proponents of FAPP, such as Zurek, they retreat into the middle of the diagram in Figure 2.8. Now, what do I mean by 'the middle of the diagram'?

I have divided the 'serious' people into different categories. There are those who believe that **U** is the whole story – that you have to take unitary evolution as the whole story. This leads to the *many-worlds* view. In this view, the cat is indeed both alive and dead but the two cats, in some sense, inhabit different universes. I shall say a bit more about that later. I have indicated some of those who have espoused this general kind of viewpoint, at least at some stage of their thinking. The many-world supporters are the ones in the middle of my diagram!

The people whom I regard as being *really serious* about $|\psi\rangle$, and I include myself among them, are those who believe that both U and R are real phenomena. Not only does unitary evolution take place out there, so long as the system is in some sense small, but there is also something different going on out there which is essentially what I have called R – it may not be exactly R but something like it which is going on out there. If you believe that then it seems that you can take one of two points of view. You could take the viewpoint that there are no new physical effects to be taken into account, and I have included the de Broglie/Bohm viewpoint here, as well as the quite different ones of Griffiths, Gell-Mann, Hartle and Omnés. R has some role to play, in addition to standard U quantum mechanics, but one would not expect to find any new effects. Then, there are those who hold the second 'really serious' viewpoint, to which I personally subscribe, that something new will have to come in and change the structure of quantum mechanics. R really does contradict U – something new is coming in. I have included the names of some of those who take this point of view at the bottom right.

I want to say something a little bit more detailed about the mathematics and specifically to look at how different viewpoints deal with Schrödinger's cat. We go back to the picture of Schrödinger's cat but now include the weightings by the complex numbers w and z (Figure 2.9(a)). The photon is split between the two states and, if you are serious about quantum mechanics, you believe the state vector is real and then you also believe that the cat must indeed be in some sort of superposition of states of being

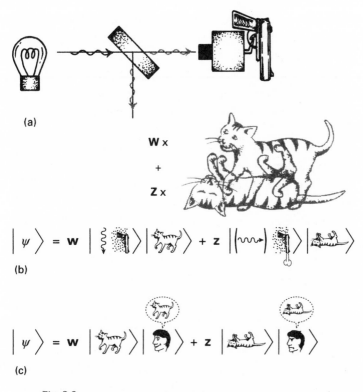

(a)

(b)

(c)

Fig. 2.9.

both dead and alive. It is very convenient to represent these states of being dead and alive using Dirac brackets, as I have shown in Figure 2.9(b). You can put cats as well as symbols inside the Dirac brackets! The cat is not the whole story because there is also the gun and the photon and the surrounding air, so there is an environment as well – each component of the state is really the product of all these effects put together, but you still have a superposition (Figure 2.9(b)).

How does the *many-worlds* point of view come to terms with this? In this, someone comes and looks at the cat and you ask:

'Why doesn't the person see these superpositions of cat states?'
Well, a many-worlds believer would describe the situation as
shown in Figure 2.9(c). There is one state of a live cat, accompanied by the person seeing and perceiving a live cat; and there
is another of the dead cat, accompanied by a person observing a
dead cat. These two alternatives are superposed: I have placed
inside the Dirac brackets the states of mind as well of the person
observing the cat in each of these two states – the person's expression reflects the state of mind of the individual. So, the view
of the many-worlds believer is that all is well – there are different
copies of the person perceiving the cat, but they inhabit 'different
universes'. You might imagine that you are one of these copies,
but there is another copy of you in another 'parallel' universe who
sees the other possibility. Of course, this is not a very economi-
cal description of the Universe but I think things are rather worse
than that for the many-worlds description. It is not just its lack of
economy that worries me. The main trouble is that it does not really solve the problem. For example, why does our consciousness
not allow us to perceive macroscopic superpositions? Let us take
the special case in which w and z are equal. Then, you can rewrite
this state as shown in Figure 2.10, that is, live cat plus dead cat together with person perceiving live cat plus person perceiving dead
cat, *plus* live cat minus dead cat together with person perceiving
live cat minus person perceiving dead cat – it is just a piece of
algebra. Now, you may say, 'Well, you cannot do that; that's not
what perception states are like!' But why not? We do not know
what perceiving means. How do we know that a state of perception could not be perceiving a live and dead cat at the same time?
Unless you know what perception is, and have a good theory of
why such mixed perception states are forbidden – and that would
be going far beyond Chapter 3 – it seems to me that this provides
no explanation. It does not explain why the perception of one or

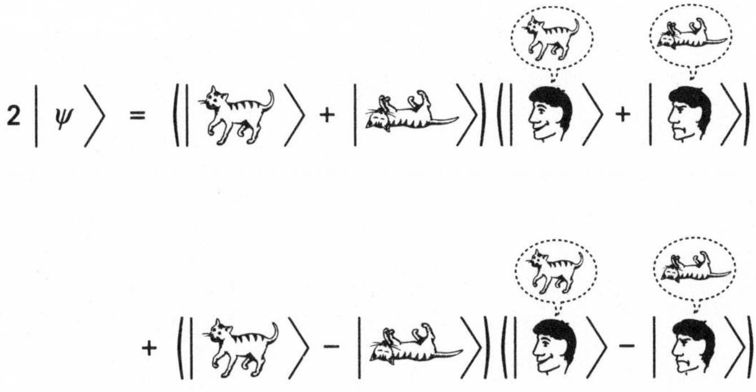

Fig. 2.10.

the other takes place but not the perception of a superposition. It could be made into a theory but you would have to have a theory of perception as well. There is another objection which is that, if we allow the numbers w and z to be general numbers, it does not tell us why the probabilities are the probabilities which come out of quantum mechanics, arrived at by the squared modulus rule I described earlier. These probabilities are, after all, things which can be tested very precisely.

Let me go a little bit further into the quantum measurement issue. I shall need to say something more about *quantum entanglement*. In Figure 2.11, I have given a description of the EPR experiment in the Bohm version, which we recall is one of the quantum Z-mysteries. How do we describe the state of the spin-$\frac{1}{2}$ particles going off in the two directions? The total spin is zero, and so, if we receive a particle with spin-up here, we know that the particle over there must have spin-down. In that case, the quantum state for the combined system would be a product of 'up-here' with 'down-there'. But, if we find that the spin is down here, then it must be up there. (These alternatives would arise if we choose to examine the spin of the particle here in the up/down

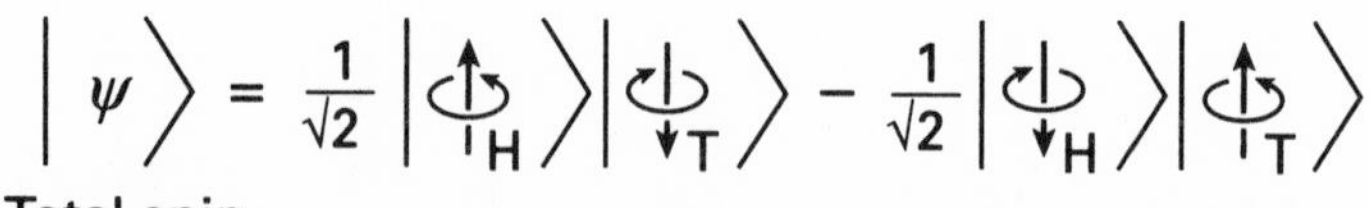

$$\left| \psi \right\rangle = \frac{1}{\sqrt{2}} \left| \overset{\uparrow}{\underset{\mathsf{H}}{\circlearrowleft}} \right\rangle \left| \underset{\downarrow\mathsf{T}}{\circlearrowleft} \right\rangle - \frac{1}{\sqrt{2}} \left| \underset{\downarrow\mathsf{H}}{\circlearrowleft} \right\rangle \left| \overset{\uparrow}{\underset{\mathsf{T}}{\circlearrowleft}} \right\rangle$$

Total spin

Fig. 2.11.

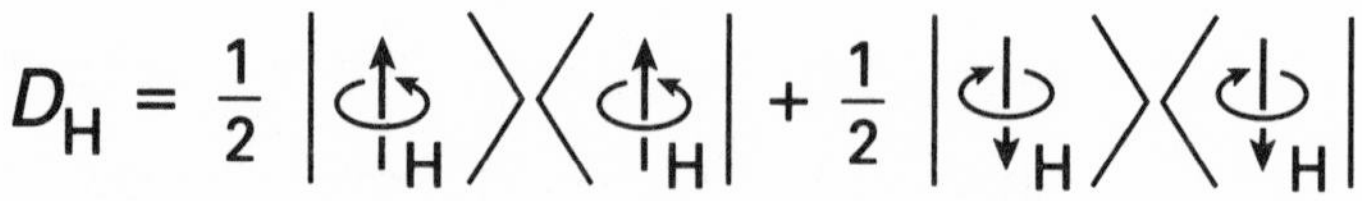

$$D_{\mathsf{H}} = \frac{1}{2} \left| \overset{\uparrow}{\underset{\mathsf{H}}{\circlearrowleft}} \right\rangle \left\langle \overset{\uparrow}{\underset{\mathsf{H}}{\circlearrowleft}} \right| + \frac{1}{2} \left| \underset{\downarrow\mathsf{H}}{\circlearrowleft} \right\rangle \left\langle \underset{\downarrow\mathsf{H}}{\circlearrowleft} \right|$$

Fig. 2.12.

direction.) To get the quantum state for the entire system, we must superimpose these alternatives. In fact, we need a minus sign to make the total spin of the pair of particles together add up to zero whichever direction we choose.

Now, suppose that we are contemplating performing a spin measurement on the particle coming towards my detector 'here' and we suppose that the other one is going off a very long way, say, to the Moon – so 'there' is on the Moon! Now, imagine that I have a colleague on the Moon who measures his particle in an up/down direction. He has an equal probability of finding his particle having spin-up or spin-down. If he finds spin-up, then the spin state of my particle must be down. If he finds spin-down, then my particle's state is up. Thus, I consider that the state vector for the particle I am about to measure is an equal probability mixture of states with spin-up and spin-down.

There is a procedure, in quantum mechanics, for handling probability mixtures like this. One uses a quantity called a *density matrix*. The density matrix that 'I over here' would use in the present situation would be the expression indicated in Figure 2.12. The first '$\frac{1}{2}$' in the expression is the probability that I find the spin over here to be upwards, and the second '$\frac{1}{2}$' in the expression is

78

the probability that I find the spin over here to be downwards. These are just ordinary classical probabilities, expressing my uncertainty about the actual spin state of the particle that I am about to measure. Ordinary probabilities are just ordinary real numbers (lying between 0 and 1), and the combination indicated in Figure 2.12 is not a quantum superposition, in which the coefficients would be complex numbers, but a probability-weighted combination. Note that the quantities that the two probability factors (of $\frac{1}{2}$) multiply are expressions which involve a first bracket factor, in which the angled bracket points to the right - called a (Dirac) *ket* vector - and also a second bracket factor in which the angled bracket points to the left - a *bra* vector. (The bra vector is what is referred to as the 'complex conjugate' of the ket vector.)

This is not the appropriate place to attempt to explain, in any detail, the nature of the mathematics that is involved in the construction of density matrices. It suffices to say that the density matrix contains all the information needed to calculate the probabilities of the results of measurements that one might perform on one part of the quantum state of the system, where it is assumed that no information is accessible concerning the remaining part of that state. In our example, the entire quantum state refers to the *pair* of particles together (an entangled state) and we assume that no information is available to me 'here' concerning measurements that might be performed 'there', on the Moon, on the partner of the particle that I am about to examine 'here'.

Now, let us change the situation slightly and suppose that my colleague on the Moon chooses to measure the spin of his particle in a left/right direction rather than up/down. For that eventuality, it is more convenient to use the description of the state given in Figure 2.13. In fact, it is exactly the same state as before, depicted in Figure 2.11, as a little algebra, based on the geometry of Figure 2.4 will show, but the state is represented differently. We still do

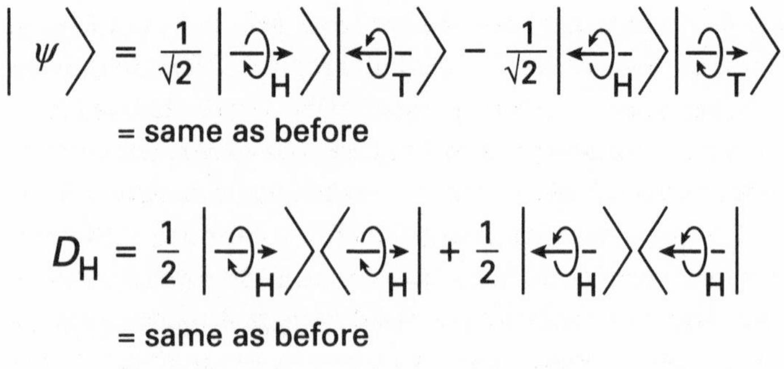

$$\left| \, \psi \, \right\rangle = \frac{1}{\sqrt{2}} \left| \, \overset{\rightarrow}{\underset{H}{\ominus}} \, \right\rangle \left| \, \overset{\leftarrow}{\underset{T}{\ominus}} \, \right\rangle - \frac{1}{\sqrt{2}} \left| \, \overset{\leftarrow}{\underset{H}{\ominus}} \, \right\rangle \left| \, \overset{\rightarrow}{\underset{T}{\ominus}} \, \right\rangle$$

= same as before

$$D_{\mathrm{H}} = \frac{1}{2} \left| \, \overset{\rightarrow}{\underset{H}{\ominus}} \, \right\rangle \left\langle \, \overset{\rightarrow}{\underset{H}{\ominus}} \, \right| + \frac{1}{2} \left| \, \overset{\leftarrow}{\underset{H}{\ominus}} \, \right\rangle \left\langle \, \overset{\leftarrow}{\underset{H}{\ominus}} \, \right|$$

= same as before

Fig. 2.13.

not know what result my colleague on the Moon will get for his (left/right) spin measurement, but we know that the probability is $\frac{1}{2}$ that he finds spin-left – in which case I must find spin-right – and $\frac{1}{2}$ that he finds spin-right – in which case I must find spin-left. Accordingly, the density matrix D_{H} must be given as in Figure 2.13, and it must turn out that this is the same density matrix as before (as given in Figure 2.12). Of course, this is as it should be. The very choice of measurement that my colleague on the Moon adopts should make no difference to what probabilities I obtain for my own measurements. (If they could make a difference, then it would be possible for my colleague to signal to me from the Moon faster than light, his message being encoded in his choice of direction of spin measurement.)

You can also check the algebra directly to see that the density matrices are indeed the same. If you know about this kind of algebra, you will know what I am talking about – if not, don't worry. The density matrix is the best you can do, if there is some part of the state which cannot be accessed. The density matrix uses probabilities in the ordinary sense, but combined with the quantum mechanical description in which there are quantum mechanical probabilities implicitly involved. If I have no knowledge about

$$\big| \, \psi \, \big\rangle = \mathbf{w} \, \big| \, \text{🐱} \big\rangle \big| \, \text{🔵} \big\rangle + \mathbf{z} \, \big| \, \text{🐱} \big\rangle \big| \, \text{⚪} \big\rangle$$

Fig. 2.14.

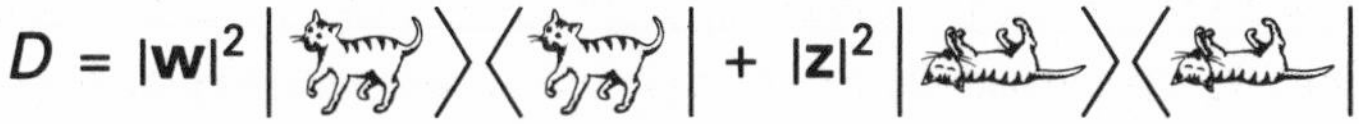

$$D = |\mathbf{w}|^2 \, \big| \, \text{🐱} \big\rangle \big\langle \, \text{🐱} \big| + |\mathbf{z}|^2 \, \big| \, \text{🐱} \big\rangle \big\langle \, \text{🐱} \big|$$

Fig. 2.15.

what is going on over 'there', this would be the best description of the state 'here' that I can give.

However, it is hard to take the position that the density matrix describes *reality*. The trouble is that I do not know that I might not, sometime later, get a message from the Moon, telling me that my colleague actually measured the state and found the answer to be such and such. Then, I know what my particle's state must *actually* be. The density matrix did not tell me *everything* about the state of my particle. For that, I really need to know the actual state of the combined pair. So, the density matrix is a sort of provisional description, and that is why it is sometimes called FAPP (i.e., for all practical purposes).

The density matrix is not usually used to describe situations like this but rather to describe situations like that shown in Figure 2.14, where, rather than having an entangled state divided between what is accessible to me 'here' and to my colleague 'there' on the Moon, the 'here' state is a cat, either dead or alive, and the 'there' state (perhaps even entirely in the same room) provides the total environment state which goes along with the cat. So, I can have live cat together with some environment plus dead cat together with another environment for the complete entangled state vector. What the FAPP people say is that you can never get enough information about the environment and therefore you don't use the state vector – you have to use the density matrix (Figure 2.15).

The density matrix then behaves like a probability mixture and the FAPP people say that, for all practical purposes, the cat is either dead or alive. This might be all right, 'for all practical purposes', but it does not give you a picture of reality – it does not tell you what might happen if some very clever person came along later and told you how to extract the information out of the environment. Somehow, it is a temporary viewpoint – good enough so long as nobody is ever be able to get that information. However, we can carry out the same analysis for the cat as we carried out for the particle in the EPR experiment. We showed that it is just as good to use the spin-left and spin-right states as to use spin-up and spin-down. We can obtain these left and right states by combining the up and down states according to the rules of quantum mechanics and get the same total entangled state vector for the pair of particles, as represented in Figure 2.13(a), and the same density matrix, as represented in Figure 2.13(b).

In the case of the cat and its environment (in the situation when the two amplitudes w and z are equal), we can do the same piece of mathematics where 'cat alive plus cat dead' plays the role of 'spin-right' and where 'cat alive minus cat dead' plays the role of 'spin-left'. We get the same state as before (Figure 2.14 with $w = z$) and the same density matrix as before (Figure 2.15 with $w = z$). Is an alive plus dead cat or an alive minus dead cat just as good as an alive cat or a dead cat? Well, that is not so obvious. But the mathematics is straightforward. There would still be the same density matrix for the cat as before (Figure 2.16). So, knowing what the density matrix is, does not help us to determine that the cat is actually either alive or dead. In other words, the cat's aliveness or deadness is not contained in the density matrix – we need more.

Not only does none of this explain why the cat is actually alive or dead (and not some combination), it does not even explain why

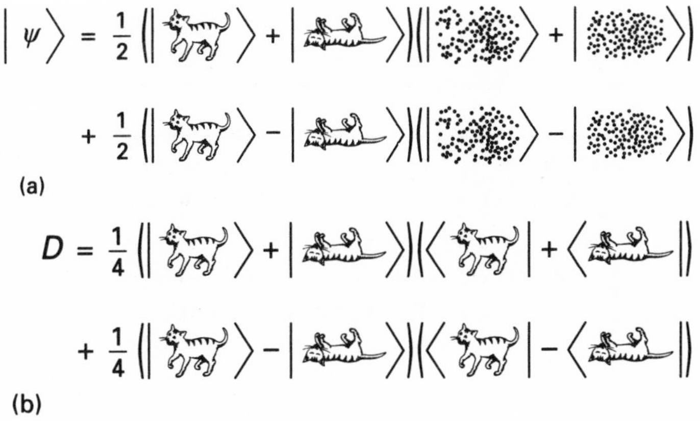

$$\left|\,\psi\,\right\rangle = \tfrac{1}{2}\left(\left|\,\vcenter{\hbox{🐱}}\,\right\rangle + \left|\,\vcenter{\hbox{🐱}}\,\right\rangle\right)\left(\left|\,\vcenter{\hbox{⬚}}\,\right\rangle + \left|\,\vcenter{\hbox{⬚}}\,\right\rangle\right)$$

$$+ \tfrac{1}{2}\left(\left|\,\vcenter{\hbox{🐱}}\,\right\rangle - \left|\,\vcenter{\hbox{🐱}}\,\right\rangle\right)\left(\left|\,\vcenter{\hbox{⬚}}\,\right\rangle - \left|\,\vcenter{\hbox{⬚}}\,\right\rangle\right)$$

(a)

$$D = \tfrac{1}{4}\left(\left|\,\vcenter{\hbox{🐱}}\,\right\rangle + \left|\,\vcenter{\hbox{🐱}}\,\right\rangle\right)\left(\left\langle\,\vcenter{\hbox{🐱}}\,\right| + \left\langle\,\vcenter{\hbox{🐱}}\,\right|\right)$$

$$+ \tfrac{1}{4}\left(\left|\,\vcenter{\hbox{🐱}}\,\right\rangle - \left|\,\vcenter{\hbox{🐱}}\,\right\rangle\right)\left(\left\langle\,\vcenter{\hbox{🐱}}\,\right| - \left\langle\,\vcenter{\hbox{🐱}}\,\right|\right)$$

(b)

Fig. 2.16.

the cat is perceived as either alive or dead. Moreover, in the case of general amplitudes, w, z, it does not explain why the relative probabilities are $|w|^2$ and $|z|^2$. My own viewpoint is that this is not good enough. I return to the diagram showing the whole of physics, but now amended to show what I think physics will have to do in the future (Figure 2.17). The procedure which I have described by the letter **R** is an approximation to something which we do not yet have. What we do not have is a thing which I call **OR** standing for *Objective Reduction*. It is an objective thing – either one thing *or* the other happens objectively. It is a missing theory. **OR** is a nice acronym because it also stands for 'or', and that is indeed what happens, one **OR** the other.

But when does this process takes place? The viewpoint I am advocating is that something goes wrong with the superposition principle when it applies to significantly differing *space-time geometries*. We encountered the idea of space-time geometries in Chapter 1 and I represent two of them in Figure 2.18(a). Further-

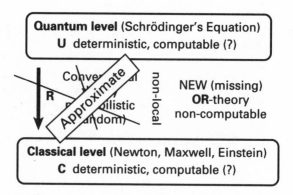

Fig. 2.17.

more, I have represented the superposition of these two space-time geometries in the figure, exactly as we did for the superposition of particles and photons. When you feel you are forced to consider superpositions of different space-times, lots of problems arise because the light cones of the two space-times can point in different directions. This is one of the big problems that people run into when they try really seriously to quantize General Relativity. Trying to do physics *within* such a funny kind of superposed space-time is something which has, in my opinion, defeated everyone so far.

What I am saying is that there are good reasons why this has defeated everybody – because this is not what one should be doing. Somehow this superposition actually becomes one **OR** the other and it is at the level of space-time that this happens (Figure 2.18(b)). Now, you might say 'This is all very well in principle, but when you try to combine quantum mechanics and general relativity, you come up with these ridiculous numbers, the Planck time and the Planck length, which are many orders of magnitude less than the normal sorts of lengths and times we deal with even in

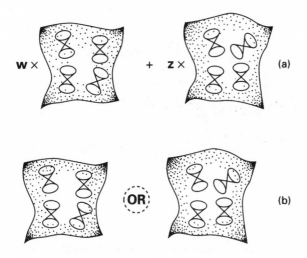

w × [diagram] + **z** × [diagram] (a)

[diagram] ⟨OR⟩ [diagram] (b)

Fig. 2.18.

particle physics. It's got nothing to do with things on the scales of cats or people. So, what can quantum gravity have to do with it?' I believe it has a lot to do with it because of the fundamental nature of what is going on.

What is the relevance of the Planck length, 10^{-33} cm, to quantum state reduction? Figure 2.19 is a highly schematic picture of a space-time which is trying to bifurcate. There is a situation leading to a superposition of two space-times, one of which could represent the dead cat and the other the live cat, and somehow these two different space-times would appear to need to be superposed. We need to ask, 'When are they going to be sufficiently different that we might worry that we ought to change the rules?' You look to see when, in some appropriate sense, the difference between these geometries is of the order of the Planck length. When the geometries start to differ by that amount you have to worry what to do and it is then that the rules might change. I should em-

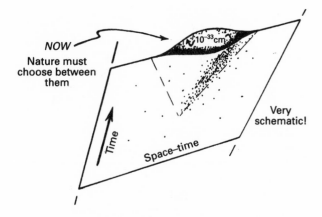

Fig. 2.19. What is the relevance of the Planck scale of
10^{-33} cm to quantum state reduction? Rough idea: when
there is sufficient mass movement between the two states
under superposition such that the two resulting
space-times differ by something of the order of 10^{-33} cm.

phasise that we are dealing with space-times here and not just
spaces. For a 'Planck scale space-time separation', a small spatial
separation corresponds to a longer time and a larger spatial separation to a shorter time. What we need is a criterion to enable
us to estimate when two space-times differ significantly and this
will lead to a *time-scale* for Nature's choice between them. Thus,
the viewpoint is that Nature chooses one or the other according
to some rule we do not understand yet.

How long does Nature take to make this choice? We can compute that time-scale in certain clear-cut situations, when the Newtonian approximation to Einstein's theory will suffice, and when
there is a clearly defined difference between the two gravitational
fields that are being subjected to quantum superposition (the two
complex amplitudes involved being about equal in magnitude).

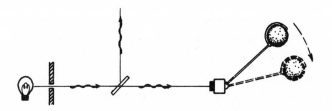

Fig. 2.20. Instead of having a cat, the measurement could consist of the simple movement of a spherical lump. How big or massive must the lump be; how far must it move; how long can the superposition last before **R** takes place?

The answer that I am suggesting is as follows. I am going to replace the cat by a lump – the cat has had a lot of work to do and it deserves a rest. How big is the lump, how far does it have to move, and what is the resulting time-scale for state-vector collapse to occur (Figure 2.20)? I am going to regard the superposition of the one state plus the other as an unstable state – it is a bit like a decaying particle or a uranium nucleus or something like that, where it might decay into one thing or another and there is a certain time-scale associated with that decay. It is a hypothesis that it is unstable, but this instability is to be an implication of the physics we do not understand. To work out the time-scale, consider the energy E which it would take to displace one instance of the lump away from the gravitational field of the other. Then, you take $\hbar$, Planck's constant divided by 2π, and divide that by this gravitational energy, and that is to be the time-scale T for the decay in this situation.

$$T = \frac{\hbar}{E}.$$

There are many schemes which follow this general type of reasoning – the general gravitational schemes all have something of this same flavour, although in detail they may differ.

There are other reasons for believing that a gravitational scheme of this nature might be a good thing to consider. One of these is that all of the other explicit schemes for quantum state reduction, which attempt to solve the quantum measurement problem by introducing some new physical phenomena, run into problems with the conservation of energy. You find that the normal rules of energy conservation tend to be violated. Maybe this is indeed the case. But, if you take a gravitational scheme, it seems to me that there is an excellent chance that we may be able to avoid that problem altogether. Although I do not know how to do this in detail, let me say what I have in mind.

In general relativity, mass and energy are rather strange things. First of all, mass is equal to energy (divided by the speed of light squared) and therefore gravitational potential energy contributes (negatively) to the mass. Accordingly, if you have two lumps which are far apart, the whole system has a slightly larger mass than if they are close together (Figure 2.21). Although the mass-energy densities (as measured by the energy-momentum tensor) are only non-zero within the lumps themselves, and the amount in each does not depend significantly on the presence of the other lump, there is a difference between the *total* energies in the two cases illustrated in Figure 2.21. The total energy is a non-local thing. There is, indeed, something fundamentally non-local about energy in General Relativity. This is certainly the case in the famous example of the binary pulsar, which I mentioned in the Chapter 1: gravitational waves carry away positive energy and mass from the system but this energy resides non-locally throughout space. Gravitational energy is an elusive thing. It seems to me that, if we had the right way of combining General Relativity with quantum mechanics, there would be a good chance of getting round the energy difficulties that plague theories of state-vector collapse. The thing is that, in the superposed state, you have to

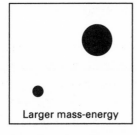
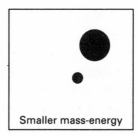

Fig. 2.21. The total mass-energy of a gravitating system involves purely gravitational contributions that are not localizable.

take into account the gravitational contribution to the energy in the superposition. But you cannot really make local sense of the energy due to gravity and so there is a basic uncertainty in the gravitational energy and that uncertainty is of the order of the energy E described above. That is just the sort of thing which one gets with unstable particles. An unstable particle has an uncertainty in its mass-energy which is related to its lifetime by this same formula.

Let me finish by examining the explicit time-scales that arise in the approach I am promoting – I shall return to this in the Chapter 3. What are the decay times for real systems in which these space-time superpositions take place? For a proton (provisionally considered to be a rigid sphere), the time-scale is a few million years. That is good, because we know from interferometer experiments with single particles that we do not see this type of thing happening. So, this is consistent. If one took a water speck with radius, say, 10^{-5} cm, the decay time would be a few hours; if it were a micron in radius, the decay time would be a twentieth of a second and, if as much as a thousandth of a centimetre, it would take about a millionth of a second. These figures indicate the sorts of scales over which this type of physics might become important.

There is, however, an essential additional ingredient which I have to bring in here. Perhaps I was slightly making fun of the FAPP point of view, but one element of that picture has to taken very seriously - that is the environment. The environment is absolutely vital in these considerations and I have ignored it in my discussion so far. You therefore have to do something much more involved. You have to consider not just the lump here superposed with the lump there but the lump with its environment superposed with the other lump and its environment. You have to look carefully to see whether the major effect is in the disturbance of the environment or in the movement of the lump. If it is in the environment, the effect is going to be random and you will not get anything different from the standard procedures. If the system can be sufficiently isolated that the environment is not involved, you might see something different from standard quantum mechanics. It would be very interesting to know if plausible experiments could be suggested - and I know of various tentative possibilities - which could test whether this type of scheme is true of nature or whether conventional quantum mechanics survives again and you really have to consider that these lumps - or even cats - must persist in such superposed states.

Let me try to summarise in Figure 2.22 what it is we have been trying to do. In this picture, I have placed the various theories at the corners of a distorted cube. The three axes of the cube correspond to the three most basic constants of physics: the gravitational constant G (horizontal axis), the speed of light taken in reciprocal form c^{-1} (diagonal axis), and the Dirac–Planck constant $\hbar$ (vertical axis downwards). Each of these constants is very tiny in ordinary terms and can be taken to be zero to a good approximation. If we take all three to be zero, we get what I am call-

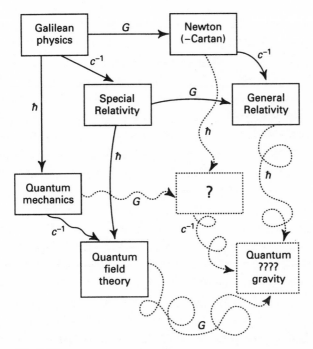

Fig. 2.22.

ing Galilean physics (top left). Including a non-zero gravitational constant moves us along horizontally to Newtonian gravitational theory (whose geometric space-time formulation was given much later by Cartan). If, instead, we allow c^{-1} to be non-zero, we get the Poincaré–Einstein–Minkowski theory of Special Relativity. The top 'square' of our distorted cube is completed if we allow both constants to be non-zero, and Einstein's General Theory of relativity is obtained. However, this generalisation is by no means straightforward – and I have illustrated this fact in Figure 2.22 by the distortions in the topmost square. Allowing $\hbar$ to be non-zero but, for the moment, reverting to $G = c^{-1} = 0$, we get standard quantum mechanics. By a not altogether direct generalisation,

c^{-1} can be incorporated also and quantum field theory is thereby obtained. This completes the left-hand face of the cube, the slight distortions indicating the lack of directness.

You might think that all we have to do now is to complete the cube and we would know the whole thing. However, it turns out that the principles of gravitational physics are in fundamental conflict with those of quantum mechanics. This shows up even with Newtonian gravity (where we maintain $c^{-1} = 0$) when we use the appropriate (Cartan) geometrical framework, in which *Einstein's Equivalence Principle* (according to which constant gravitational fields are indistinguishable from accelerations) is used. This was pointed out to me by Joy Christian, who also provided the inspiration behind my Figure 2.22. As yet, no appropriate union between quantum mechanics and Newtonian gravity – which fully takes into account Einstein's Equivalence Principle, as is done in the classical theory by Cartan's geometry – has yet come to light. In my own clear opinion, this union would have to accommodate the phenomenon of *quantum state reduction* – roughly along the lines of the **OR** ideas outlined earlier in this chapter. Such a union would clearly be very far from straightforward in its completion of the back face of the cube in Figure 2.22. The full theory, incorporating all three constants, $\hbar$, G and c^{-1}, in which the whole 'cube' is completed, would have to be something even more subtle and mathematically sophisticated. This is clearly a matter for the future.

CHAPTER 3

Physics and the Mind

The first two chapters were concerned with the physical world and the mathematical rules we use to describe it, how remarkably accurate they are and how strange they sometimes appear to be. In this third chapter, I shall talk about the *mental world* and, in particular, how it is related to the physical world. I suppose that Bishop Berkeley would have thought that in some sense the physical world emerges out of our mental world, whereas the more usual scientific viewpoint is that somehow mentality is a feature of some kind of physical structure.

Popper introduced a third world called the *World of Culture* (Figure 3.1). He viewed this world as a product of mentality and so he had a hierarchy of worlds as illustrated in Figure 3.2. In this picture, the mental world is, in some way, related to (emergent from?) the physical world and, somehow, culture arises out of mentality.

Now, I want to look at things a little bit differently. Rather than thinking, as Popper did, of culture as arising out of our mentality, I prefer to believe that the worlds are connected as shown in Figure 3.3. Moreover, my 'World III' is not really the World of Culture but the world of Platonic absolutes – particularly absolute mathematical truth. In this way, the arrangement of Figure 1.3, illustrating the profound dependence of the physi-

Fig. 3.1. 'World III' of Karl Popper.

cal world on precise mathematical laws, is incorporated into our figure.

Much of this chapter will concern the relationship between all these different worlds. It seems to me that there is a fundamental problem with the idea that mentality arises out of physicality – this is something which philosophers worry about for very good reasons. The things we talk about in physics are matter, physical things, massive objects, particles, space, time, energy and so on. How could our feelings, our perception of redness, or of happiness have anything to do with physics? I regard that as a mystery. We can regard the arrows connecting the different worlds in Figure 3.3 as mysteries. In the first two chapters, I discussed the relation between mathematics and physics (Mystery 1). I referred to Wigner's remarks concerning this relation. He regarded it as

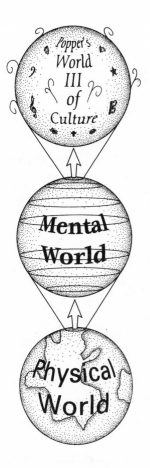

Fig. 3.2.

very extraordinary and I do too. Why is it that the physical world seems to obey mathematical laws in such an extremely precise way? Not only that, but the mathematics which seems to be in control of our physical world is exceptionally fruitful and powerful, simply *as* mathematics. I regard this relationship as a deep mystery.

95

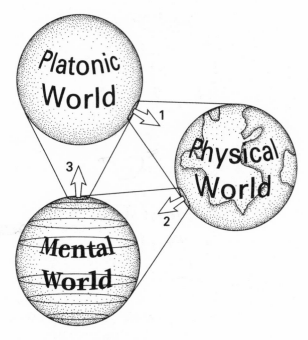

Fig. 3.3. Three Worlds and three Mysteries.

In this chapter, I shall examine Mystery 2: the mystery of the relation of the physical world to the world of mentality. But in relation to this, we shall also have to consider Mystery 3: what underlies our ability to access mathematical truth? When I referred to the Platonic world in the first two chapters, I was primarily talking about mathematics and the mathematical concepts one has to call upon to describe the physical world. One has the feeling that the mathematics needed to describe these things is out there. There is also, however, the common feeling that these mathematical constructions are products of our mentality, that is, mathematics is a product of the human mind. One can look at things in this way but it is not really the mathematician's way of looking at mathematical truth; nor is it my way of looking at it either. So although, there is an arrow joining the mental world and the Platonic world, I do not mean to indicate that this, or indeed any of these arrows, implies that any of these worlds simply emerges out of any of the others. Although there may be a sense in which they are emerging, the arrows are simply meant to represent the fact that there is a relationship between the different worlds.

More important is the fact that Figure 3.3 represents three prejudices of mine. One of them is that that the entire physical world can, in principle, be described in terms of mathematics. I am not saying that all of mathematics can be used to describe physics. What I am saying is that, if you choose the right bits of mathematics, these describe the physical world very accurately and so the physical world behaves according to mathematics. Thus, there is a small part of the Platonic world which encompasses our physical world. Likewise, I am also not saying that everything in the physical world has mentality. Rather, I am suggesting that there are not mental objects floating around out there which are not based in physicality. This is my second prejudice. There is a third prejudice that, in our understanding of mathematics, in principle at least, any individual item in the Platonic world is accessible by our mentality, in some sense. Some people might well worry about this third prejudice – indeed they may worry about all three prejudices. I should say that it was only after I had drawn this diagram that I realised that it reflected these three prejudices of mine. I shall come back to this diagram at the end of the chapter.

Let me now say something about *human consciousness*. In particular, is this a question we should think about in terms of scientific explanation? My own viewpoint is very much that we should. In particular, I take the arrow joining the physical and the mental worlds very seriously. In other words, we have the challenge of understanding the mental world in terms of the physical world.

I have summarised some characteristics of the physical and the mental worlds in Figure 3.4. On the right-hand side, we have aspects of the *physical world* – it is perceived as being governed by precise mathematical, physical laws, as discussed in the first two chapters. On the left-hand side, we have consciousness which belongs to the *mental world* and words like 'soul', 'spirit', 'religion', and so on, are frequently incurred. Nowadays, people prefer scientific explanations for things. Moreover, they tend to think that you could, in principle, put any scientific description on a computer; accordingly, if you have a mathematical description of something, you should, in principle, be able to put it on a computer. This is something which I shall *argue strongly against* in this chapter, despite my physicalist bias.

The terms used to describe the physical laws in Figure 3.4 are *predictive, calculational* – these have to do with whether or not we have *determinism* in our physical laws and whether or not we could use a computer to simulate the action of these laws. On the one hand, there is the view that mental things like emotion, aesthetics, creativity, inspiration and art are examples of things which it would be hard to see emerging from some kind of calculational description. At the other 'scientific' extreme, some people would say, 'We are just computers; we may not know how to describe these things yet, but somehow, if we knew the right kinds of computations to carry out, we would be able to describe all the mental things listed in Figure 3.4'. The word *emergence* is often used to describe this process. These qualities 'emerge', according to these people, as a result of the right kind of computational activity.

What is *consciousness?* Well, I don't know how to define it. I think this is not the moment to attempt to define consciousness,

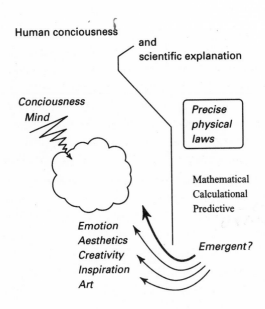

Fig. 3.4.

since we do not know what it is. I believe that it is a physically accessible concept; yet, to define it would probably be to define the wrong thing. I am, however, going to describe it, to some degree. It seems to me that there are at least two different aspects to consciousness. On the one hand, there are *passive* manifestations of consciousness, which involve *awareness*. I use this category to include things like perceptions of colour, of harmonies, the use of memory, and so on. On the other hand, there are its *active* manifestations, which involve concepts like free will and the carrying out of actions under our free will. The use of such terms reflects different aspects of our consciousness.

I shall concentrate here mainly on something else which involves consciousness in an essential way. It is different from both passive and active aspects of consciousness, and perhaps is something somewhere in between. I refer to the use of the term *understanding*, or perhaps *insight*, which is often a better word. I am not going to define these terms either – I don't know what they mean. There are two other words I do not understand – *awareness* and *intelligence*. Well, why am I talking about things when I do not know what they really mean? It is probably because I am a mathematician and mathematicians do not mind so much about that sort of thing. They do not need precise definitions of the things they are taking about, provided they can say something about the *connections* between them. The first key point here is that it seems to me that intelligence is something which requires understanding. To use the term intelligence in a context in which we deny that any understanding is present seems to me to be unreasonable. Likewise, understanding without any awareness is also a bit of a nonsense. Understanding requires some sort of awareness. That is the second key point. So, that means that intelligence requires awareness. Although I am not defining any of these terms, it seems to me to be reasonable to insist upon these relations between them.

There are various viewpoints which one can take about the relationship between conscious thinking and computation. I have summarised in Table 3.1 four approaches to awareness, which I have labelled **A, B, C** and **D**.

The viewpoint I am calling **A**, which is sometimes called *strong artificial intelligence* (strong AI) or (computational) *functionalism*, asserts that all thinking is simply the carrying out of some computation and therefore, if you carry out the appropriate computations, awareness will be the result.

I have labelled the second viewpoint **B** and, according to it, you could, in principle, simulate the action of a brain, when its owner is aware of something. The difference between **A** and **B** is that, although that activity can be simulated, that mere simulation would

Table 3.1.

A	All thinking is computation; in particular, feelings of conscious awareness are evoked merely by the carrying out of appropriate computations.
B	Awareness is a feature of the brain's physical action; and, whereas any physical action can be simulated computationally, computational simulation cannot by itself evoke awareness.
C	Appropriate physical action of the brain evokes awareness, but this physical action cannot even be properly simulated computationally.
D	Awareness cannot be explained by physical, computational, or any other scientific terms.

not of itself, according to **B**, have any feelings or any awareness – there is something else going on, which is perhaps to do with the physical construction of the object. So a brain made up of neurons and so on would be allowed to be aware, whereas a simulation of the activity of that brain would not be aware. This is, as far as I can make out, the point of view which has been promoted by John Searle.

Next, there is my own point of view, which I have called **C**. According to this view, in agreement with **B**, there is something in the physical action of the brain which evokes awareness – in other words, it is something in the physics we have to turn to, but this physical action is something which cannot even be simulated computationally. There is no simulation which you could carry out of that action. This requires that there should be something in the physical action of the brain which is beyond computation.

Finally, there is always viewpoint **D**, according to which it is a mistake to look at these issues in terms of science at all. Perhaps awareness cannot be explained in scientific terms.

I am very much a proponent of viewpoint **C**. There are, however, various varieties of **C**. There is what might be called **weak C** and **strong C**. **Weak C** is the viewpoint that somehow, in known physics, you would only need to look carefully enough and you would find certain types of action which are beyond computation. When I say 'beyond computation', I have to be a little bit more explicit, as I shall in just a moment. According to **weak C**, there is nothing we need look for outside known physics in order to find the appropriate non-computational action. **Strong C**, in contrast, requires that there should be something outside known physics; our physical understanding is inappropriate for the description of awareness. It is incomplete, and, as you will have gathered from Chapter 2, I do indeed believe that our physical picture is incomplete, as I indicated by Figure 2.17. From the viewpoint of **strong C**, maybe future science will explain the nature of consciousness but present day science does not.

I included some words in Figure 2.17 which I did not comment upon at the time, in particular, the word *computable*. In the standard picture, one has basically computable physics at the quantum level, and the classical level is probably computable, although there are technical questions about how one goes from computable discrete systems to continuous systems. It is an important point but let me not worry about it here. In fact, it seems to me that proponents of **weak C** would have to find something in these uncertainties, something which cannot be explained in terms of a computable description.

To get from the quantum to the classical level in the conventional picture, we introduce the procedure which I have called **R** and which is an entirely probabilistic action. What we have then is computability together with randomness. I am going to argue that this is not good enough – we need something different and this new theory, which bridges these two levels, has to be a non-computable theory. I shall say a little bit more about what I mean by that term in a moment.

So, this is my version of **strong C**: we look for the non-computability in the physics which bridges the quantum and the classical levels. This is quite a tall order. I am saying that not only do we need new physics, but we also we need new physics which is relevant to the action of the brain.

First of all, let us address the question as to whether or not it is plausible that there is something beyond computation in our understanding. Let me give you a very nice example of a simple chess problem. Computers play chess very well nowadays. However, when the chess problem shown in Figure 3.5 was given to the most powerful computer available at that time, the Deep Thought computer, it did a very stupid thing. In this chess position, the white pieces are heavily outnumbered by the black – there are two extra black rooks and a black bishop. This should be an enormous advantage, were it not for the fact that there is a barrier of pawns, which boxes in all the black pieces. So, all white has to do is to wander around behind its barrier of white pawns and there is no chance of losing the game. However, when this position was given to Deep Thought, it instantly gobbled up the black rook, opened up the barrier of pawns and obtained a hopelessly lost position. The reason it did that is that it had been programmed to compute move, after move, after move, after move ... to a certain depth and then to count up the pieces, or something like that. In this instance, that was not good enough. Of course, if it went move, after move, after move a few more times, it might have been able to do it. The thing is that chess is a computational game. In this case, the human player sees the barrier of pawns and understands that it is impenetrable. The computer did not have that understanding – it simply computed move after move. So, this example is an illustration

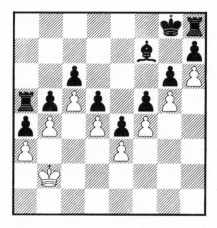

Fig. 3.5. White to play and draw – easy for humans, but Deep Thought took the rook! (Problem by William Hartston, taken from an article by Jane Seymore and David Norwood in *New Scientist*, No. 1889, page 23, 1993).

of the difference between mere computation and the quality of understanding.

Here is another example (Figure 3.6). There is a great temptation to take the black rook with the white bishop but the correct thing to do is to pretend that the white bishop is a pawn and use it to create another barrier of pawns. Once you have taught the computer to recognise barriers of pawns, it might be able to solve the first problem but it would fail on the second because it needs an extra level of understanding. However, you might think that with enough care, it would be possible to program in all possible levels of understanding. Well, maybe you could with chess. The trouble is that chess is a computational game and so ultimately it would be possible to compute every possibility, to the very end, with a powerful enough computer. This is far beyond the capacity of present day computers but it would, in principle, be possible.

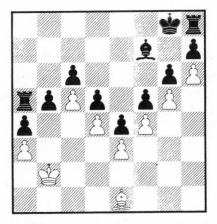

Fig. 3.6. White to play and draw – again easy enough for humans, but a normal expert chess computer will take the rook (from a Turing test by William Hartston and David Norwood).

Nonetheless, one has the feeling that there is something else going on with 'understanding' other than direct computation. Certainly, the way we approach these chess problems is very different from what a computer does.

Can we make a stronger argument that there is indeed something in our understanding which is different from computation? Well, we can. I don't want to spend too much time on this argument, although it is actually the foundation stone of the whole discussion. But I do have to spend a little time on it, even though the argument can get a bit technical. The first 200 pages of *Shadows of the Mind* were devoted to trying to show that there are no loop-holes in the argument that I am about to give you.

Let me say something about *computations*. Computations are what a computer does. Real computers have a limited amount of storage capacity but I am going to consider an idealised computer, called a *Turing machine*, which differs from an ordinary general purpose computer only in the fact that it has an unlimited amount of storage space and can go on computing forever without making any mistakes and without ever wearing out. Let me give an example of a computation. A computation need not just involve doing arithmetic but can involve carrying out logical operations as well. Here is an example:

• *Find a number that is not the sum of three square numbers.*

By a number, I mean a *natural number* such as 0, 1, 2, 3, 4, 5, ... and by 'squares numbers' I mean the numbers 0^2, 1^2, 2^2, 3^2, 4^2, 5^2, ... Here is how you might do it – it is a pretty stupid way of doing it in practice, but it does illustrate what we can mean by a computation. We start with 0 and test whether it is the sum of three square numbers. You look at all the squares which are less than or equal to 0 and there is only 0^2. Therefore, we can only try

$$0 = 0^2 + 0^2 + 0^2$$

which turns out to be true, and so 0 is the sum of three squares. Next, we try 1. We write down all the possible ways of adding up all the numbers whose squares are less than or equal to one and see if we can sum three of them to make 1. Well, we can:

$$1 = 0^2 + 0^2 + 1^2$$

We can carry on in this rather tedious way, as indicated in Table 3.2, until we reach the number 7, where you can see that there is no way in which you can add together three squares of 0^2, 1^2 and 2^2 in any combination to make the number 7 – all the possibilities are shown in the table. Thus, 7 is the answer – it is the smallest number which is not the sum of three square numbers. This is an example of a computation.

In that example, we were lucky, because the computation came to an end, whereas there are certain computations which do not

Table 3.2.

Try 0	squares ≤ 0 are	0^2	$0 = 0^2 + 0^2 + 0^2$
Try 1	squares ≤ 1 are	$0^2, 1^2$	$1 = 0^2 + 0^2 + 1^2$
Try 2	squares ≤ 2 are	$0^2, 1^2$	$2 = 0^2 + 1^2 + 1^2$
Try 3	squares ≤ 3 are	$0^2, 1^2$	$3 = 1^2 + 1^2 + 1^2$
Try 4	squares ≤ 4 are	$0^2, 1^2, 2^2$	$4 = 0^2 + 0^2 + 2^2$
Try 5	squares ≤ 5 are	$0^2, 1^2, 2^2$	$5 = 0^2 + 1^2 + 2^2$
Try 6	squares ≤ 6 are	$0^2, 1^2, 2^2$	$6 = 1^2 + 1^2 + 2^2$
Try 7	squares ≤ 7 are	$0^2, 1^2, 2^2$	$7 \neq 0^2 + 0^2 + 0^2$
			$7 \neq 0^2 + 0^2 + 1^2$
			$7 \neq 0^2 + 0^2 + 2^2$
			$7 \neq 0^2 + 1^2 + 1^2$
			$7 \neq 0^2 + 1^2 + 2^2$
			$7 \neq 0^2 + 2^2 + 2^2$
			$7 \neq 1^2 + 1^2 + 1^2$
			$7 \neq 1^2 + 1^2 + 2^2$
			$7 \neq 1^2 + 2^2 + 2^2$
			$7 \neq 2^2 + 2^2 + 2^2$

actually terminate at all. For example, suppose I change the problem slightly:

- *Find a number that is not the sum of four square numbers.*

There is a famous theorem due to the eighteenth century mathematican Lagrange who proved that every number can be expressed as the sum of four squares. So, if you simply went on in a mindless way to find such a number, the computer would simply chug away forever and never find any answer. This illustrates the fact that there are indeed some computations which do not terminate.

Lagrange's Theorem is quite tricky to prove and so here is another easier one which I hope everyone can appreciate!

- *Find an odd number that is the sum of two even numbers.*

You could set your computer to do that and it would go on end-lessly because we know that when we add two even numbers to-gether we always get an even number.

Here is a distinctly trickier example:

- *Find a even number greater than 2 that is not the sum of two primes.*

Does this computation ever terminate? It is generally believed that it does not, but this is a mere conjecture, known as the Goldbach Conjecture, and it is so difficult that nobody knows for sure whether or not it is true. So here are (probably) three non-stopping calculations, an easy one, a hard one and a third which is so hard that no one yet knows whether it actually stops or not.

Let us now ask the question:

- Are mathematicians using some computational algorithm (say *A*) to convince themselves that certain computations do not terminate?

For example, did Lagrange have some sort of computer prog-ramme in his head, which ultimately led him to the conclusion that every number is the sum of four squares? You do not even need to be Lagrange – you simply have to be someone who can fol-low Lagrange's argument. Notice that I am not concerned about the issue of originality, only about the question of understanding. That is why I have expressed the question in the above way – to 'convince themselves' means to create understanding.

The technical term for a statement of the nature of those that we have just been considering is that it is a Π_1-*sentence*. A Π_1-*sentence* is an assertion that some specified computation does not terminate. To appreciate the argument which follows, we need only think about sentences of this nature. I want to convince you that there is no such algorithm *A*.

In order to do that, I need to generalise slightly. I have to talk about computations which depend upon a natural number n. Here are some examples:

- *Find a natural number that is not the sum of n square numbers.*

We have seen from Lagrange's Theorem, that if n is four or more, it does not stop. But if n is up to three, then it does stop. The next computation is:

- *Find an odd number that is the sum of n even numbers*

Well, it does not matter what n is – that is not going to help you at all. It does not stop for any value of n whatever. For the extension of the Goldbach Conjecture, we have:

- *Find an even number greater than 2 that is not the sum of up to n prime numbers.*

If Goldbach's Conjecture is true, then this computation will stop for no n whatever (other than 0 and 1). In a sense, the larger n is, the easier this is. In fact, I believe that there is a large enough value of n for which it is actually known that the computation is 'non-stopping'.

The important point is that these types of computation depend upon the natural number n. This is, in effect, central to the famous argument known as the *Gödel Argument*. I shall discuss it in a form due to Alan Turing, but I shall use his argument in a slightly different way from the way he did. If you do not like mathematical arguments, you may switch off for a little bit. The result is the important thing. But, in any case, the argument is not very complicated – only confusing!

Computations, which act upon a number n, are basically computer programs. You can make a list of computer programs and attach a number, say p, to each of them. So you feed your general-purpose computer some number p and it chugs away, carrying out that 'pth' computation as applied to whichever number n you have selected. The number p is written as a suffix in our notation. So, I list these computer programs, or computations, which act upon the number n, one after another.

$$C_0(n), C_1(n), C_2(n), C_3(n), \ldots, C_p(n), \ldots$$

We are going to suppose that this is a list of *all* possible computations $C_p(n)$ and that we can find some effective way of ordering these computer programs, so that the number p labels the pth program in the ordering. Then, $C_p(n)$ stands for the pth program applied to the natural number n.

Now, let us suppose that we have some computational, or algorithmic, procedure A, which can act upon a pair of numbers (p, n), and when that procedure comes to an end, it provides us with a valid demonstration that the computation $C_p(n)$ does not terminate. A will not necessarily always work, in the sense that there may be some computations $C_p(n)$ that are non-terminating, whereas $A(p, n)$ does not terminate either. But I want to insist that A does not actually make mistakes and so, if $A(p, n)$ does terminate, $C_p(n)$ actually does not. Let us try to imagine that human mathematicians act according to some computational procedure A when they formulate (or follow) some rigorous mathematical demonstration of a mathematical proposition (say, of a Π_1-sentence). Suppose that they are also allowed to *know* what A is and that they *believe* it to be a sound procedure. We are going to try to imagine that A encapsulates *all* the procedures available to human mathematicians for convincingly demonstrating that computations do not stop. The procedure A starts by looking at the letter p to select the computer programme and then it looks at the number n, to find out which number it is to act upon. Then, if the computational procedure A comes to an end, that implies that the computation $C_p(n)$ does not end. Thus,

If $A(p, n)$ stops, then $C_p(n)$ does not stop.   (1)

This is what A's job is – it provides the way of unassailably convincing oneself that certain computations do not terminate.

Now, suppose we put $p = n$. This may seem a funny thing to do. It is the famous procedure known as *Cantor's Diagonal Procedure* and there is nothing wrong with using it. Then, we reach the conclusion that

If $A(n, n)$ stops, then $C_n(n)$ doesn't stop.

But now, $A(n, n)$ depends only upon one number and so $A(n, n)$ must be one of the computer programs $C_p(n)$, because that list is exhaustive for computations acting on a single variable n. Let us suppose that the computer program which is identical to $A(n, n)$ is labelled k. Then,

$$A(n, n) = C_k(n).$$

Now, we put $n = k$ and we find that

$$A(k, k) = C_k(k).$$

Then, we look at the statement (1) and conclude that

If $A(k, k)$ stops, then $C_k(k)$ does not stop.

But $A(k, k)$ is the same as $C_k(k)$. Therefore, if $C_k(k)$ stops, it doesn't stop. That means it doesn't stop. That is pretty clear logic. But here is the catch – this particular computation does not stop, and if we believe in A, then we must also believe that $C_k(k)$ does not stop. But A also does not stop, and therefore it does not 'know' that $C_k(k)$ does not stop. Therefore, the computational

111

procedure cannot, after all, encompass the totality of mathematical reasoning for deciding that certain computations do not stop – that is, for establishing the truth of Π_1-sentences. That is the gist of the Gödel–Turing argument in the form that I need.

You may question the full force of this argument. What it clearly says is that mathematical insight cannot be coded in the form of some computation that *we can know to be correct*. People sometimes argue about this but that seems to me to be its clear implication. It is interesting to read what Turing and Gödel said about this result. Here is Turing's statement:

> In other words, then, if a machine is expected to be
> infallible, it cannot also be intelligent. There are several
> theorems which say almost exactly that. But these
> theorems say nothing about how much intelligence may be
> displayed if a machine makes no pretence at infallibility.

So, his idea was that Gödel–Turing-type arguments can be reconciled with the idea that mathematicians are essentially computers if the algorithmic procedures they act according to, in order to ascertain mathematical truth, are basically *unsound*. We can restrict attention to arithmetical statements, for example Π_1-sentences, which form a pretty restrictive type of statement. I believe that Turing actually thought that the human mind does use algorithms but that these algorithms are just wrong – that is, they are indeed unsound. I find this a rather implausible standpoint, particularly because one is not concerned here with how one might get inspiration, but simply how one might follow an argument and understand it. It seems to me that Turing's position is not very plausible. According to my scheme, Turing would have been an **A** person.

Let us see what Gödel said. In my scheme, he was a **D** person. Thus, even although Turing and Gödel had the same evidence in front of them, they came to essentially opposite conclusions. Nevertheless, although Gödel did not really believe that mathematical insight could be reduced to computation, he was not able to rule out this possibility rigorously. Here is what Gödel said:

> On the other hand, on the basis of what has been proved so far, it remains possible that there may exist (and even be empirically discovered) a theorem-proving machine which in fact *is* equivalent to mathematical intuition, but cannot be *proved* to be so, nor can be proved to yield only *correct* theorems of finitary number theory.

His argument was that there is a 'loophole' to the direct use of the Gödel–Turing argument as a refutation of computationalism (or, functionalism), namely that mathematicians might be using an algorithmic procedure which is sound but which we cannot know for sure is sound. So it was the *knowable* part which Gödel thought was a loophole and the *sound* part which Turing settled on.

My own view is that neither of these is plausibly the way out of the argument. What the Gödel–Turing theorem says is that if any algorithmic procedure (for establishing Π_1-sentences) is found to be sound, then you can immediately exhibit something which goes outside it. It could be that we are using an algorithmic procedure which cannot be known to be sound and there might be some sort of learning device which enables us to develop this facility. These topics, and many others, are dealt with *ad nauseam* in my book *Shadows of the Mind*. I do not want to go into these ramifications here. I shall only mention two points.

How could this putative algorithm have arisen? In the case of human beings, it presumably would have to have come about through natural selection, or, in the case of robots, it would have to be created by deliberate AI (artificial intelligence) construction.

Fig. 3.7. For our remote ancestors, a specific ability to do sophisticated mathematics can hardly have been a selective advantage, but a general ability to *understand* could well have.

I shall not go into these arguments in detail but simply illustrate them by two cartoons from my book.

The first cartoon has to do with *natural selection* (Figure 3.7). You can see the mathematician who is not in a very happy position from the point of view of natural selection, because you can see that there is a sabre-toothed tiger about to pounce upon him. In contrast, his cousins in the other part of the cartoon are catching mammoths, building houses, growing crops and so on. These things involve understanding but are not specific to mathematics. Thus, the quality of understanding could be the thing we were selected for but specific algorithms for doing mathematics could not really have been.

The other cartoon has to do with *deliberate AI construction* and there is a little story in my book about an AI expert of the future having a discussion with the robot (Figure 3.8). The complete argument given in the book is somewhat long and complicated – I don't feel it is really necessary to go into it all here. My original

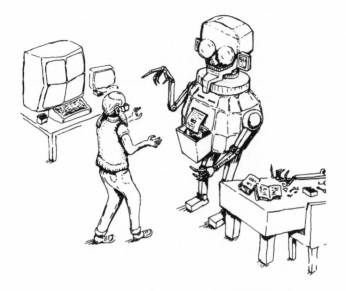

Fig. 3.8. Albert Imperator confronts the Mathematically
Justified Cybersystem. In *Shadows of the Mind* the first
200 pages are devoted to disposing of criticisms of the
use of the Gödel-Turing argument. The essence of these
new arguments is contained in the dialogue between the
AI person and his robot.

use of the Gödel–Turing argument had been attacked by all sorts
of people from all sorts of different angles and all these different
points had to be addressed. I tried to encapsulate most of these
new arguments that are presented in *Shadows* in the discussion
which the AI person has with his robot.

Let us come back to the question of what is going on. Gödel's
argument concerns particular statements about numbers. What
Gödel tells us is that no system of computational rules can char-
acterise the properties of the *natural numbers*. Despite the fact
that there is no computational way of characterising the natural
numbers, any child knows what they are. All you do is to show

115

the child different numbers of objects, as illustrated in Figure 3.9, and after a while they can abstract the notion of natural number from these particular instances of it. You do not give the child a set of computational rules – what you are doing is enabling the child to 'understand' what natural numbers are. I would say that the child is able to make some kind of 'contact' with the Platonic world of mathematics. Some people do not like this way of talking about mathematical insight but, nevertheless, it seems to me that one has to take some kind of view of this nature as to what is going on. Somehow, the natural numbers are already 'there', existing somewhere in the Platonic world and we have access to that world through our ability to be aware of things. If we were simply mindless computers, we would not have that ability. Rules are not the things which enable us to comprehend the nature of natural numbers, as Gödel's Theorem shows. Understanding what the natural numbers 'are' is a good instance of Platonic contact.

Thus, I am saying that, more generally, mathematical understanding is not a computational thing, but something quite different, depending upon our ability to be aware of things. Some people might say, 'Well, all you claim to have proved is that mathematical insight is not computational. That doesn't say much about other forms of consciousness.' But it seems to me that this is good enough. It is unreasonable to draw a line between mathematical understanding and any other kind of understanding. That is what I was trying to illustrate with my first cartoon (Figure 3.7). Understanding is something which is not specific to mathematics. Human beings develop this quality of general understanding and it is *not* a computational quality because mathematical understanding is not. Nor do I draw a line between human understanding and human consciousness generally. So, although I said I do not know what human consciousness is, it seems to me that human understanding is an instance of it, or at least is something

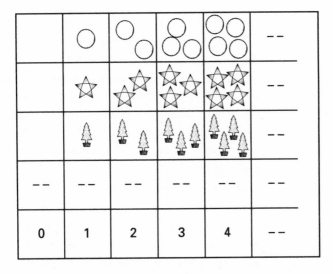

Fig. 3.9. The Platonic notion of a natural number can be abstracted by a child, from just a few simple examples.

which requires it. I am also not going to draw a line between human consciousness and animal consciousness either. I might get into trouble with different sets of people here. It seems to me that humans are very much like many other kinds of animal and, although we may have a bit better understanding of things than some of our cousins, nevertheless, they also have some kind of understanding, and so also must have awareness.

Therefore, non-computability in *some* aspect of consciousness and, specifically, in mathematical understanding, strongly suggests that non-computability should be a feature of *all* consciousness. This is my suggestion.

Now, what do I mean by non-computability? I have talked a lot about it but I should give an example of something which is non-computational to show what I mean. What I am about to describe to you is an example of what is often called a *toy model universe* – it is the sort of thing that physicists do when they cannot think of anything better to do. (It is actually not such a bad thing to do!) The point about a toy model is that it does not purport to be an actual model of the Universe. It may reflect certain features of the Universe but it is not meant to be taken seriously as a model for the actual Universe. This particular toy model is certainly not meant to be taken seriously in that sense. It is presented merely to illustrate a certain point.

In this model, there is a discrete time which runs 0, 1, 2, 3, 4, ... and the state of the Universe at any one time is to be given by a *polyomino set*. What is a polyomino set? Well, some examples are illustrated in Figure 3.10. A polyomino is a collection of squares all stuck together along various edges to form some plane shape. I am concerned with sets of polyominoes. Now, in this toy model, the state of the universe at any one moment is to be given by two separate finite sets of polyominoes. In Figure 3.10, I envisage a complete list of all possible finite sets of polyominoes, listed S_0, S_1, S_2, ..., in some computational way. What is the evolution, or dynamics, of this ridiculous universe? We start at time zero with the polyomino sets (S_0, S_0) and then we continue with other pairs of polyomino sets according to a certain precise rule. This rule depends on whether or not it is possible to use a given polyomino set to tile the whole plane simply using the polyominoes of that set. The question is therefore, can you cover the entire plane without gaps or overlaps using only the polyominoes of the given set. Now, suppose that the universe state of the toy model at one instant of time is the pair of polyomino sets (S_q, S_r). The rule for the evolution of this model is that, if you can tile the plane with the polyominoes of S_q, then you go on to the next one S_{q+1}, giving the pair (S_{q+1}, S_r) at the next instant of time. If you cannot, then, in addition, you must swap the pair around to give (S_r, S_{q+1}). It

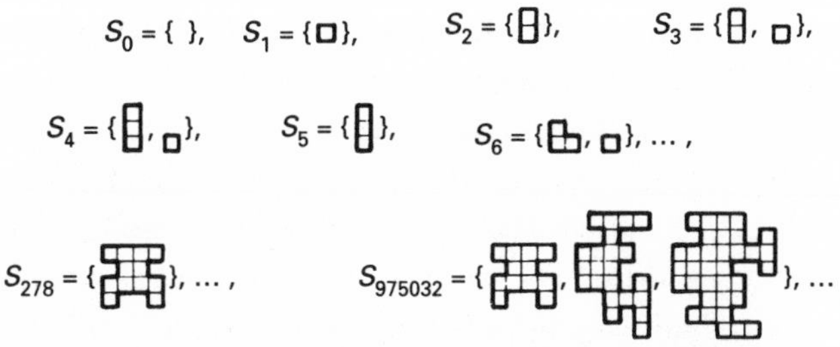

$$S_0 = \{ \ \}, \quad S_1 = \{\square\}, \quad S_2 = \{\boxminus\}, \quad S_3 = \{\boxminus, \square\},$$

$$S_4 = \{\boxminus, \square\}, \quad S_5 = \{\boxminus\}, \quad S_6 = \{\boxminus, \square\}, \dots ,$$

$$S_{278} = \{\boxplus\}, \dots , \quad S_{975032} = \{\boxplus, \boxplus, \boxplus\}, \dots$$

Fig. 3.10. A non-computable toy model universe. The different states of this deterministic but non-computable toy universe are given in terms of pairs of finite sets of polyominoes. So long as the first set of the pair tiles the plane, the time-evolution proceeds with the first set increasing in numerical order and the second 'marking time'. When the first set does not tile the plane, the two swap over as the evolution continues. It would go something like: (S_0, S_0), (S_0, S_1), (S_1, S_1), (S_2, S_1), (S_3, S_1), $(S_4, 1), \dots , (S_{278}, S_{251})$, (S_{251}, S_{279}), $(S_{252}, S_{279}), \dots .$

is a very simple, stupid little universe – what is the point of it? The point is that, although its evolution is entirely deterministic – I have given you a very clear, absolutely deterministic rule as to how the universe is to evolve – it is *non-computable*. It follows from a theorem of Robert Berger that there is no computer action which can simulate the evolution of this universe because there is no computational decision procedure for deciding when a polyomino set will tile the plane.

This illustrates the point that computability and determinism are different things. Some examples of polyomino tilings are shown in Figure 3.11. In examples (a) and (b), these shapes can tile a complete plane as illustrated. In example (c), the left-hand and right-hand shapes on their own cannot tile a plane – in both cases, they leave gaps. But, taken together, they can tile the entire plane, as illustrated in (c). Example (d) will also tile the plane – it can only tile the plane in the way shown and this illustrates how complicated these tilings can become.

Things can get worse, however. Let me show you the example of Figure 3.12 – in fact, Robert Berger's theorem depends upon the existence of tile sets like this. The three tiles shown at the top of the figure will cover the entire plane, but there is no way of doing this in such a way that the pattern repeats itself. It is always different as you keep on going out and it is not so easy to see that you can actually do it. But, nevertheless, it can be done and the existence of tilings like this goes into Robert Berger's argument from which it follows that there is no computer programme which can simulate this toy universe.

What about the actual Universe? Well, I have argued in the Chapter 2 that there is something fundamental missing in our physics. Is there any reason from physics itself to think that there might be something non-computable in this missing physics? Well, I think there is some such reason to believe this – that the true quantum gravity theory might be non-computable. The idea is not entirely plucked out of the air. I shall indicate that non-computability is a feature of two independent approaches to quantum gravity. What is distinctive about these particular approaches is that they involve the quantum superposition of four-dimensional space-times. Many other approaches involve only superpositions of three-dimensional spaces.

The first is the Geroch–Hartle scheme for quantum gravity, which turns out to have a non-computable element, because it invokes a result, due to Markov, which asserts that topological 4-manifolds are not computationally classifiable. I shall not go into this technical matter, but it does show that this feature of non-computability has already come up in a natural way in attempts to combine General Relativity and quantum mechanics.

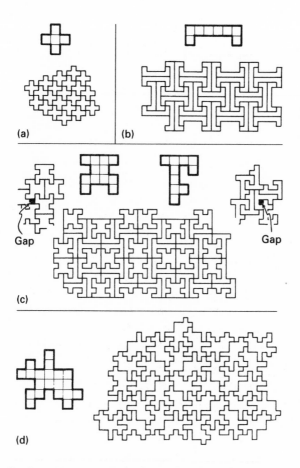

Fig. 3.11. Various sets of polyominoes that will tile the infinite Euclidean plane (reflected tiles being allowed). Neither of the polyominoes in set (c), if taken by itself, will tile the plane, however.

Fig. 3.12. This set of three polyominoes will tile the plane only non-periodically.

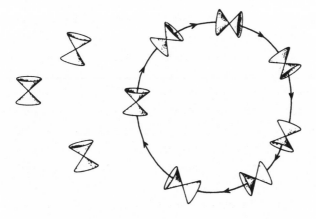

Fig. 3.13. With severe enough light-cone tilting in a
space–time, closed timelike lines can occur.

The second place where non-computability has come up in an
approach to quantum gravity is in the work of David Deutsch. It
appeared in a preprint he issued and then, irritatingly, when the
paper appeared in print, the argument was nowhere to be found!
I asked him about this and he assured me that he took it out, not
because it was wrong, but because it was not relevant to the rest
of the paper. His point of view is that, in these funny superpositions of space-times, you have to consider at least the possibility
that some of these potential universes might have closed timelike lines (Figure 3.13). In these, causality has gone all crazy, the
future and the past get mixed up and causal influences go round
in loops. Now, although these need only play a role as counterfactuals, as in the bomb-testing problem of Chapter 2, they still
have an influence on what actually happens. I would not say that
this is a clear argument, but it is at least an indication that there
could easily be something of a non-computational nature in the
correct theory, if we ever find it.

I want to raise another issue. I stressed that determinism and computability are different things. It does slightly bear upon the issue of *free will*. In philosophical discussions, free will has always been talked about in terms of determinism. In other words, 'Is our future determined by our past?' and issues of that nature. It seems to me that there are lots of other questions which might be asked. For example, 'Is the future determined *computably* by the past?' – that is a different question.

These considerations raise all sorts of other issues. I shall only raise them – I certainly shan't try to answer them. There are always great arguments about the extent to which our actions are determined by *our heredity* and *our environment*. Something which strangely is not often mentioned is the role of *chance elements*. In some sense, all of these things are beyond our control. You might ask the question: 'Is there something else, perhaps a thing called the *self*, which is different from all of these and which lies beyond such influences?' Even legal issues have relevance to such an idea. For example, the questions of rights or responsibilities seem to depend upon the actions of an independent 'self'. It may be quite a subtle issue. First, there is the relatively straight-forward issue of *determinism* and *non-determinism*. The normal kind of non-determinism just involves random elements, but that does not help you very much. These chance elements are still beyond our control. You might have *non-computability* instead. You might have *higher-order types of non-computability*. Indeed, it is a curious thing that the Gödel-type arguments which I have given can actually be applied at different levels. They can be applied at the level of what Turing called *oracle machines* – the argument is actually much more general than the way I presented it above. So one has to consider the question of whether or not there might be some kind of higher-order type of non-computability involved in the way that the actual Universe evolves. Perhaps our feelings of free will have something to do with this.

I have talked about contact with some sort of Platonic world – what is the nature of this 'Platonic contact'? There are some types of words which would seem to involve non-computable elements – for example, judgement, common sense, insight, aesthetic sensibility, compassion, morality, ... These seem to me to be things which are not just features of computation. Up to this point, I have talked about the Platonic world primarily in terms of mathematics, but there are other things which one might also include. Plato would certainly argue that not only the true, but also the good and the beautiful are absolute (Platonic) concepts. If there indeed exists some sort of contact with Platonic absolutes which our awareness enables us to achieve, and which cannot be explained in terms of computational behaviour, then that seems to me to be an important issue.

Well, what about our brains? Figure 3.14 shows a little bit of a brain. A major constituent of the brain is its system of *neurons*. An important part of each neuron is a very long fibre known as its *axon*. The axons bifucrate into separate strands at various places and each of these finally terminates at a thing called a *synapse*. These synapses are the junctions where signals are transferred from each neuron to (mainly) other neurons by means of chemical substances called neurotransmitters. Some synapses are excitatory in nature, with neurotransmitters which tend to enhance the firing of the next neuron and others are inhibitory, tending to suppress the firing of the next neuron. We can refer to the reliability of a synapse in passing the message from one neutron to the next as the *strength* of the synapse. If the synapses all had fixed strengths, the brain would be very much like a computer. However, it is certainly the case that these synaptic strengths can change and there are various theories about how they change. For

125

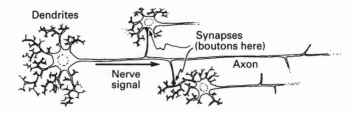

Fig. 3.14. A sketch of a neuron, connected to some others via synapses.

example, the Hebb mechanism was one of the earliest suggestions for this process. The point is, however, that all the mechanisms for inducing changes which have been suggested are of a computational nature, albeit with additional probabilistic elements. So, if you have some kind of computational-probabilistic rule which tells you how these strengths change, then you could still simulate the action of the system of neurons and synapses by a computer (since probabilistic elements can easily be computationally simulated also) and we obtain the type of system illustrated in Figure 3.15.

The units illustrated in Figure 3.15, which we may imagine to be transistors, could play the role of the neurons in the brain. For example, we can consider specific electronic devices known as *artificial neural networks*. In these networks, various rules are incorporated concerning how the synapse strengths change, usually in order to improve the quality of some output. But the rules are always of a computational nature. It is easy to see that this must be so, for the very good reason that people simulate these things on computers. That is the test. If you are able to put the model on a computer, then it is computable. For example, Gerald Edelman has some suggestions about how the brain might work which he claims are not computational. What does he do? He has

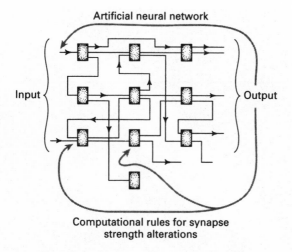

Computational rules for synapse
strength alterations

Fig. 3.15.

a computer which simulates all these suggestions. So, if there is
a computer which is supposed to simulate it, then it is computational.

I want to address the question, 'What are individual neurons
doing? Are they just acting as computational units?' Well, neurons are cells and cells are very elaborate things. In fact, they are
so elaborate that, even if you only had one of them, you could
still do very complicated things. For example, a paramecium, a
one-celled animal, can swim towards food, retreat from danger,
negotiate obstacles and, apparently, learn by experience (Figure
3.16). These are all qualities which you would think would require
a nervous system but the paramecium certainly has no nervous
system. The best you could do would be if the paramecium were
a neuron itself! There are certainly no neurons in a paramecium
– there is only a single cell. The same sort of statement would
apply to an amoeba. The question is 'How do they do it?'

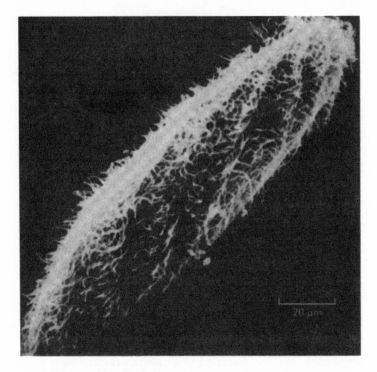

Fig. 3.16. A paramecium. Note the hair-like cilia that are
used for swimming. These form the external extremities
of the paramecium's *cytoskeleton.*

One suggestion is that the *cytoskeleton* – the structure which,
among other things, gives the cell its shape – is what is controlling
the complicated actions of these one-celled animals. In the case of
the paramecium, the little hairs, or cilia, which it uses to swim are
the ends of the cytoskeleton, and these are largely made up of the
little tubelike structures called *microtubules.* The cytoskeleton is
made up of these microtubules, as well as actin and intermediate
filaments. Amoebae also move around, effectively using microtubules to push out their pseudopods.

Microtubules are extraordinary things. The cilia, which the paramecium uses to swim about, are basically bundles of microtubules. Furthermore, microtubules are very much involved in mitosis, that is, in cell division. This is true of the microtubules in ordinary cells but not, apparently, in neurons – neurons do not divide and this may be an important difference. The control centre of the cytoskeleton is a structure known as the *centrosome*, whose most prominent part, the *centriole*, consists of two bundles of microtubules in the shape of a separated 'T'. At a critical stage, as the centrosome divides, each of the two cylinders in the centriole grows another, so as to make two centriole 'T's which then separate from each other, each giving the appearancce of dragging a bundle of microtubules with it. These microtubule fibres somehow connect the two parts of the divided centrosome to the separate DNA strands in the nucleus of the cell and the DNA strands then separate. This process initiates cell division.

This is not what happens in neurons because neurons do not divide, and so the microtubules must be doing something else. What are they doing in neurons? Well, they are probably doing lots of things, including transporting neurotransmitter molecules within the cell, but one thing in which they do seem to be involved is in determining the strengths of the synapses. In Figure 3.17, a blow-up of a neuron and synapse is shown, in which the rough locations of the microtubules as well as the actin fibres are also indicated. One way in which the strength of a synapse might be influenced by microtubules is in influencing the nature of a *dendritic spine* (Figure 3.17). Such spines occur at many synapses, and they can apparently grow or shrink or otherwise change their nature. Such changes can be induced by alterations in the actin within them, actin being an essential constituent of the mechanism of muscle contraction. Neighbouring microtubules could strongly influence this actin which, in turn, could influence the shape or

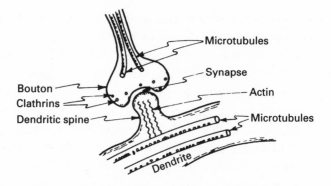

Fig. 3.17. Clathrins (and microtubule endings) inhabit the axon's synaptic boutons and seem to be involved in influencing the strength of synpases. This could occur via actin filaments in dendritic spines.

the dielectric properties of the synapse connection. There are at least two other different ways in which microtubules could be involved in influencing the strengths of synapses. They are certainly involved in transporting the neurotransmittor chemicals which transmit the signal from one neuron to the next. It is the microtubules which carry them along the axons and dendrites and so their activity would influence the concentration of these chemicals at the end of the axon and in the dendrites. This, in turn, could influence the synapse strength. Another microtubule influence would be in neuron growth and degeneration, altering the very network of neuron connections.

What are microtubules? A sketch of one of them is shown in Figure 3.18. They are little tubes made of proteins called *tubulin*. They are interesting in various ways. The tubulin proteins seem to have (at least) two different states, or conformations, and they can change from one conformation to the other. Apparently, messages can be sent along the tubes. In fact, Stuart Hameroff and his

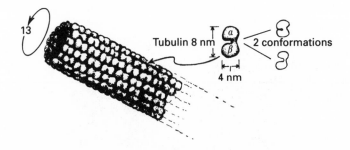

Fig. 3.18. A microtubule. It is a hollow tube, normally consisting of 13 columns of tubulin dimers. Each tubulin molecule appears to be capable of (at least) two conformations.

colleagues have interesting ideas about how signals might be sent along the tubes. According to Hameroff, the microtubules may behave like *cellular automatons* and complicated signals could be sent along them. Think of the two different conformations of each tubulin as representing the '0' and the '1' of a digital computer. Thus, a single microtubule could itself behave like a computer, and one has to take this into account if one is considering what neurons are doing. Each neuron does not just behave like a switch but rather involves many, many microtubules and each microtubule could be doing very complicated things.

This is where my own ideas make an entry. It might be that quantum mechanics is important in understanding these processes. One of the things that excites me most about microtubules is that they are *tubes*. Being tubes, there is a plausible possibility that they might be able to isolate what is going on in their interiors from the random activity in the environment. In Chapter 2, I made the claim that we need some new form of **OR** physics and, if it is going to be relevant, there must be quantum-superposed mass movements which are well isolated from the

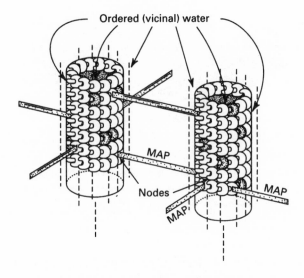

Fig. 3.19. Systems of microtubules within (collections of) neurons might sustain large-scale quantum-coherent activity, individual **OR** occurrences constituting conscious events. Effective isolation of this activity is required, possibly by ordered water surrounding the microtubules. An interconnecting system of microtubule-associated proteins (MAPs) could 'tune' this activity, attaching to the microtubules at 'nodes'.

environment. It may well be that, within the tubes, there is some kind of large-scale, quantum coherent activity, somewhat like a superconductor. Significant mass movement would be involved only when this activity begins to get coupled to the (Hameroff-type) tubulin conformations, where now the 'cellular automaton' behaviour would itself be subject to quantum superposition. The sort of thing which might take place is illustrated in Figure 3.19.

As part of this picture, there would have to be some type of co-herent quantum oscillation taking place within the tubes which would need to extend over very large areas of the brain. There were some suggestions of this general type put forward by Herbert Frölich many years ago, making it somewhat plausible that there might be things of this nature in biological systems. Microtubules seem to be a good candidate for the structures within which this large-scale quantum coherent activity might take place. When I use the term 'large-scale', you will recall that, in Chapter 2, I described the EPR puzzle and the effects of quantum non-locality, which show that effects which are widely separated cannot be regarded as separate from each other. Non-local effects like this occur in quantum mechanics and they cannot be undertood in terms of one thing being separate from another – some sort of global activity is taking place.

It seems to me that consciousness is something global. Therefore, any physical process responsible for consciousness would have to be something with an essentially global character. Quantum coherence certainly fits the bill in this respect. For such large-scale quantum coherence to be possible, we need a high degree of isolation, as might be supplied by the microtubule walls. However, we also need more, when the tubulin conformations begin to get involved. This needed additional insulation from the environment might be supplied by ordered water just outside the microtubules. Ordered water (which is known to exist in living cells) would be likely also to be an important ingredient of any quantum-coherent oscillation taking place inside the tubes. Though a tall order, perhaps it is not totally unreasonable that all this might be the case.

Quantum oscillations within the tubes would have to be coupled in some way to the action of the microtubules, namely the cellular automaton activity that Hameroff talks about, but now his idea has to be combined with quantum mechanics. Thus, we now must have not only computational activity in the ordinary sense, but also quantum computation which involves superpositions of different such actions. If that were the whole story, we would still be at the quantum level. At a certain point, the quantum state might get entangled with the environment. Then we would leap up to the classical level in a seemingly random way, in accordance with the usual **R** procedure of quantum mechanics. This is no good if we want genuine non-computability to be coming in. For that, the non-computable aspects of **OR** have to manifest themselves, and that requires excellent isolation. Thus, I claim that we need something in the brain which has enough isolation that the new **OR** physics has a chance to play a important role. What we would need is for these superposed microtubular computations, once they get going, to be sufficiently isolated that this new physics would indeed come into play.

So the picture I have is that, for a while, these quantum computations go on and they keep themselves isolated from the rest of the material for long enough – perhaps something of the order of nearly a second – that the kinds of criteria I was talking about take over from the standard quantum procedures, the non-computational ingredients come in, and we get something essentially different from standard quantum theory.

Of course, there is a good deal of speculation in many of these ideas. Yet they offer some genuine prospect of a much more specific and quantitative picture of the relation between consciousness and biophysical processes than has been available in other approaches. We can at least begin to make a computation as to how many neurons need to be involved in order that this **OR** action could possibly become relevent. What is needed is some estimate of T, the time-scale I talked about towards the end of Chapter 2. In other words, assuming that conscious events are related to such **OR** occurrences, what do we estimate T to be? How long does consciousness require? There are two types of experiments, both associated with Libet and his associates relevant to these ideas. One deals with free will, or active consciousness; the other with sensation, or passive consciousness.

First, consider free will. In Libet's and Kornhuber's experiments, a subject is asked to press a button, at a time entirely determined by his (or her) volition. Electrodes are placed on the subject's head to detect electrical activity in the brain. Many repeated trials are made and the results averaged (Figure 3.20(a)). The upshot is that there is some clear indication of such electrical activity about a full second ahead of the time that the subject believes that the actual decision is made. So free will seems to involve some kind of time delay, of the order of a second.

More remarkable are the passive experiments, which are more difficult to carry out. They seem to suggest that it takes about half a second of activity in the brain before a person becomes passively aware of something (Figure 3.20(b)). In these experiments, there are ways of blocking out the conscious experience of a stimulus to the skin, up to about half a second *after* this stimulus actually occurred! In those cases when the blocking procedure is not effected, the subject believes the experience of the skin stimulus to have occurred at the actual time of that stimulus. Yet it could have been blocked out up to half a second after the actual time of the stimulus. These are very puzzling experiments, particularly when taken together. They suggest that conscious willing appears to need about a second, conscious sensation needs about half a second. If you imagine that consciousness is something which does something, then you are presented with almost a paradox. You need half a second before you become conscious of some event. Then you try to bring your conscousness into play so as to do something with it. You need another second for your free will to do that something – that is, you need about a total of a

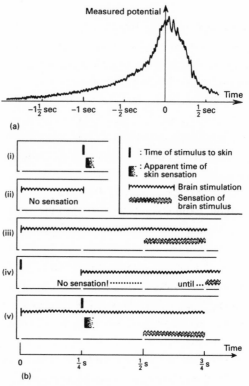

Fig. 3.20. (a) Kornhuber's experiment, later repeated and refined by Libet and his colleagues. The decision to flex the finger appears to be made at time 0, yet the precursor signal (averaged over many trials) suggest a 'foreknowledge' of the intention to flex. (b) Libet's experiment. (i) The stimulus to the skin 'seems' to be perceived at about the actual time of the stimulus. (ii) A cortical stimulus of less than half a second is not perceived. (iii) A cortical stimulus of over half a second is perceived from half a second onwards. (iv) Such a cortical stimulus can 'backwards mask' an earlier skin stimulus, indicating that awareness of the skin stimulus had actually *not yet taken place* by the time of the cortical stimulus. (v) If a skin stimulus is applied shortly *after* such a cortical stimulus, then the skin awareness is 'referred back' but the cortical awareness is not.

136

second and a half. So, if anything requires a consciously willed response, you need about a second and a half before you could actually make use of it. Well, I find that rather hard to believe. Consider ordinary conversation, for example. It seems to me that, although a good deal of conversation could well be automatic and unconscious, the fact that it takes a second and a half to make a *conscious* response seems to me to be very strange.

My way of looking at this is that there may well be something in the way that we interpret such experiments that makes some presumption that the physics we are using is basically classical physics. Remember the bomb-testing problem where we talked about counterfactuals and the fact that counterfactual events could have an influence upon things, even though they did not actually occur. The ordinary sort of logic one uses tends to go wrong if one is not careful. We have to bear in mind how quantum systems behave and so it might be that something funny is going on in these timings because of quantum non-locality and quantum counterfactuals. It is very difficult to understand quantum non-locality within the framework of Special Relativity. My own view is that, to understand quantum non-locality, we shall require a radically new theory. This new theory will not just be a slight modification of quantum mechanics but something as different from standard quantum mechanics as General Relativity is different from Newtonian gravity. It would have to be something which has a completely different conceptual framework. In this picture, quantum non-locality would be built into the theory.

In Chapter 2, non-locality was shown to be something, which although it is very puzzling, can still be described mathematically. Let me show you the picture of an impossible triangle in Figure 3.21. You may ask, 'Where is the impossibility?' Can you locate it? You can cover up various parts of the picture and, whichever bit of the triangle you cover up, the picture suddenly becomes pos-

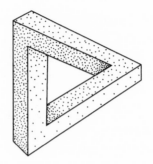

Where is the impossibility?

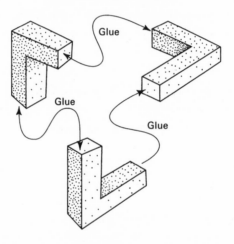

Fig. 3.21. An impossible triangle. The 'impossibility' cannot be localised; yet it can be defined in precise mathematical terms as an abstraction from the 'glueing rules' underlying its construction.

sible. So you cannot say that the impossibility is at any specific place in the picture – the impossibility is a feature of the whole structure. Nevertheless, there are precise mathematical ways in which you can talk about such things. This can be done in terms of breaking it apart, glueing it together and extracting certain abstract mathematical ideas from the detailed total pattern of glueings. The notion of *cohomology* is the appropriate notion in this case. This notion provides us with a means of calculating the degree of impossibility of this figure. This is the kind of non-local mathematics that might well be involved in our new theory.

It is not supposed to be an accident that Figure 3.21 looks like Figure 3.3! The drawing of Figure 3.3 was deliberately made so as to emphasize an element of paradox. There is something distinctly mysterious about the way that these three worlds interrelate with one another – where each seems almost to 'emerge' from a small part of its predecessor. Yet, as with Figure 3.21, with further understanding we may be able to come to terms with or even resolve some of this mystery. It is important to recognize puzzles and mysteries when they occur. But just because there is something very puzzling going on does not mean that we shall never be able to understand it.

REFERENCES

Albrecht-Buehler, G. (1981) Does the geometric design of centrioles imply their function? *Cell Motility* **1**, 237–45.

Albrecht-Buehler, G. (1991) Surface extensions of 3T3 cells towards distant infrared light sources, *J. Cell Biol.* **114**, 493–502.

Aspect, A., Grangier, P., and Roger, G. (1982). Experimental realization of Einstein–Podolsky–Rosen–Bohm *Gedankenexperiment*: a new violation of Bell's inequalities, *Phys. Rev. Lett* **48**, 91–4.

Beckenstein, J. (1972) Black holes and the second law, *Lett. Nuovo Cim.*, **4**, 737–40.

Bell, J.S. (1987) *Speakable and Unspeakable in Quantum Mechanics* (Cambridge University Press, Cambridge).

Bell, J.S. (1990) Against measurement, *Physics World* **3**, 33-40.

Berger, R. (1966) The undecidability of the domino problem, *Memoirs Amer. Math. Soc.*, No. 66 (72pp.).

Bohm, D. and Hiley, B. (1994). *The Undivided Universe*. (Routledge, London).

Davenport, H. (1968) *The Higher Arithmetic*, 3rd edn, (Hutchinson's University Library, London).

Deeke, L., Grötzinger, B., and Kornhuber, H.H. (1976). Voluntary finger movements in man: cerebral potentials and theory, *Biol. Cybernetics*, **23**, 99.

Deutch, D. (1985) Quantum theory, the Church-Turing principle and the universal quantum computer, *Proc. Roy. Soc. (Lond.)* **A400**, 97-117.

DeWitt, B.S. and Graham, R.D., eds. (1973) *The Many-Worlds Interpretation of Quantum Mechanics*. (Princeton University Press, Princeton).

Diósi, L. (1989) Models for universal reduction of macroscopic quantum fluctuations, *Phys. Rev.* **A40**, 1165-74.

Fröhlich, H. (1968). Long-range coherence and energy storage in biological systems, *Int. J. of Quantum. Chem.*, **II**, 641-9.

Gell-Mann, M. and Hartle, J.B. (1993) Classical equations for quantum systems, *Phys. Rev. D* **47**, 3345-82.

Geroch, R. and Hartle, J. (1986) Computability and physical theories, *Found. Phys.* **16**, 533.

Gödel, K. (1931) Über formal unentscheidbare Sätze der Principia Mathematica und verwandter System 1, *Monatshefte für Mathematik und Physik* **38**, 173-98.

Golomb, S.W. (1966) *Polyominoes*. (Scribner and Sons, London).

Haag, R. (1992) *Local Quantum Physics: Fields, Particles, Algebras*, (Springer-Verlag, Berlin).

Hameroff, S.R. and Penrose, R. (1996). Orchestrated reduction of quantum coherence in brain microtubules - a model for consciousness. In *Toward a Science of Consciousness: Contributions from the 1994 Tucson Conference*, eds, S. Hameroff, A. Kaszniak and A. Scott (MT Press, Cambridge MA).

Hameroff, S.R. and Penrose, R. (1996). Conscious events as orchestrated space-time selections. *J. Consciousness Studies*, **3**, 36-53.

Hameroff, S.R. and Watt, R.C. (1982). Information processing in microtubules, *J. Theor. Biol.* **98**, 549-61.

Hawking, S.W. (1975) Particle creation by black holes, *Comm. Math. Phys.* **43**, 199-220.

Hughston, L.P., Jozsa, R., and Wooters, W.K. (1993) A complete classification of quantum ensembles having a given density matrix, *Phys. Letters* **A183**, 14-18.

Károlyházy, F. (1966) Gravitation and quantum mechanics of macroscopic bodies, *Nuovo Cim.* **A42**, 390.

Károlyházy, F. (1974) Gravitation and quantum mechanics of macroscopic bodies, *Magyar Fizikai PolyoirMat* **12**, 24.

Károlyházy, F., Frenkel, A. and Lukács, B. (1986) On the possible role of gravity on the reduction of the wave function. In *Quantum Concepts in Space and Time* eds. R. Penrose and C.J. Isham (Oxford University Press, Oxford) pp. 109-28.

Kibble, T.W.B. (1981) Is a semi-classical theory of gravity viable? In *Quantum Gravity 2: A Second Oxford Symposium*, eds C.J. Isham, R. Penrose and D.W. Sciama (Oxford University Press, Oxford) pp. 63-80.

Libet, B. (1992) The neural time-factor in perception, volition and free will, *Review de Métaphysique et de Morale*, **2**, 255-72.

Libet, B., Wright, E.W. Jr, Feinstein, B. and Pearl, D.K. (1979) Subjective referral of the timing for a conscious sensory experience, *Brain*, **102**, 193-224.

Lockwood, M. (1989) *Mind, Brain and the Quantum* (Basil Blackwell, Oxford).

Lucas, J.R. (1961) Minds, Machines and Gödel, *Philsophy* **36**, 120-4; reprinted in Alan Ross Anderson (1964) *Minds and Machines* (Prentice-Hall, New Jersey).

Majorana, E. (1932) Atomi orientati in campo magnetico variabile, *Nuovo Cimento* **9**, 43-50.

Moravec, H. (1988) *Mind Children: The Future of Robot and Human Intelligence* (Harvard University Press, Cambridge, MA).

Omnés, R. (1992) Consistent interpretations of quantum mechanics, *Rev. Mod. Phys.*, **64**, 339-82.

Pearle, P. (1989) Combining stochastic dynamical state-vector reduction with spontaneous localisation, *Phys. Rev.*, **A39**, 2277-89.

Penrose, R. (1989) *The Emperor's New Mind: Concerning Computers, Minds, and the Laws of Physics*, (Oxford University Press, Oxford).

Penrose, R. (1989) Difficulties with inflationary cosmology, in *Proceedings of the 14th Texas Symposium on Relativistic Astrophysics*, ed. E. Fenves, *Annals of NY Acad. Sci.* **571**, 249 (NY Acad. Science, New York).

Penrose, R. (1991) On the cohomology of impossible figures [La cohomologie des figures impossibles], *Structural Topology [Topologie structurale]* **17**, 11-16.

Penrose, R. (1994) *Shadows of the Mind: An Approach to the Missing Science of Consciousness* (Oxford University Press, Oxford).

Penrose, R. (1996) On gravity's role in quantum state reduction, *Gen. Rel. Grav.* **28**, 581.

Percival, I.C. (1995) Quantum spacetime fluctuations and primary state diffusion, *Proc. R. Soc. Lond.* **A451**, 503-13.

Schrödinger, E. (1935) Die gegenwärtige Situation in der Quantenmechanik, *Naturwissenschaftenp*, **23**, 807-12, 823-8, 844-9. (Translation by J.T. Trimmer (1980) in *Proc. Amer. Phil. Soc.*, **124**, 323-38).

Schrödinger, E. (1935) Probability relations between separated systems, *Proc. Camb. Phil. Soc.*, **31**, 555-63.

Searle, J.R. (1980) Minds, Brains and Programs, in *The Behavioral and Brain Sciences*, Vol. 3 (Cambridge University Press, Cambridge).

Seymore, J. and Norwood, D. (1993) A game for life, *New Scientist* **139**, No. 1889, 23-6.

Squires, E. (1990) On an alleged proof of the quantum probability law *Phys. Lett.* **A145**, 67-8.

Turing, A.M. (1937) On computable numbers with an application to the Entscheidungsproblem, *Proc. Lond. Math. Soc. (ser. 2)* **42**, 230-65.; a correction **43**, 544-6.

Turing, A.M. (1939) Systems of logic based on ordinals, *P. Lond. Math. Soc.*, **45**, 161-228.

von Neumann, J. (1955) *Mathematical Foundations of Quantum Mechanics.* (Princeton University Press, Princeton).

Wigner, E.P. (1960) The unreasonable effectiveness of mathematics in the physical sciences, *Commun. Pure Appl. Math.*, **13**, 1-14.

Zurek, W.H. (1991) Decoherence and the transition from quantum to classical, *Physics Today*, **44** (No. 10), 36-44.

CHAPTER 4

On Mentality, Quantum Mechanics and the Actualization of Potentialities

ABNER SHIMONY

INTRODUCTION

What I admire most in Roger Penrose's work is the spirit of his investigations – the combination of technical expertise, daring and determination to get to the heart of the matter. He follows Hilbert's great exhortation, *'Wir müssen wissen, wir werden wissen.'*[1] As to the program of his investigation, I agree with him on three basic theses. First, mentality can be treated scientifically. Second, the ideas of quantum mechanics are relevant to the mind-body problem. Third, the quantum mechanical problem of the actualization of potentialities is a genuine physical problem that cannot be solved without modifying the quantum formalism. I am sceptical, however, of many of the details of Roger's elaboration of these three theses and hope that my criticism will stimulate him to make improvements.

4.1 THE STATUS OF MENTALITY IN NATURE

About one quarter of Chapters 1–3 and about half of his book *Shadows of the Mind* (hereafter abbreviated SM) are devoted to establishing the non-algorithmic character of human mathematical ability. Hilary Putnam's review[2] of SM claimed that there are some lacunae in the argument – that Roger neglects the possibility of a programme for a Turing machine that simulates human mathematical ability but is not provably sound, and the possibility of such a programme being so complex that in practice a human mind could not understand it. I am not convinced by Roger's answer to Putnam,[3] but on the other hand I am not sufficiently knowledgeable about proof theory to adjudicate with confidence. But it seems to me that the issue is tangential to Roger's central concern, and that he is an alpinist who has tried to climb the wrong mountain. His central thesis, that there is something about mental acts that cannot be achieved by any artificial computer, is not dependent upon establishing the non-algorithmic character of human mathematical operations. Indeed, as an adjunct to his long Gödelian argument, Roger presents (SM pp. 40-1) John Searle's 'Chinese room' argument that a correct computation by an automaton does not constitute understanding. The core of the argument is that a human subject could be trained to behave as an automaton by behaviourally following directions acoustically presented in Chinese, even though the subject does not understand Chinese and knows that this is the case. A subject who correctly carried out a computation by following these directions can directly compare the normal experience of computing by understanding and the abnormal experience of computing like an automaton. The mathematical truth established by the computation in question may be entirely trivial, and nevertheless the difference between mechanically computing and understanding is intuitively evident.

What Searle, with Roger's endorsement, has defended concerning mathematical understanding applies also to other aspects of conscious experience - to sensory qualia, to sensations of pain and pleasure, to feelings of volition, to intentionality (which is the experienced reference to objects or concepts or propositions), etc. Within the general philosophy of physicalism there are various

145

strategies for accounting for these phenomena.[4] In two-aspect theories, these experiences are regarded as aspects of specific brain states; other theories identify a mental experience with a class of brain states, the class being so subtle that an explicit physical characterization of it cannot be given, thereby precluding the explicit 'reduction' of a mental concept to physical concepts; functionalist theories identify mental experiences with formal programs that can in principle be realised by many different physical systems even though, as a contingent matter of fact, they are realized by a network of neurons. A recurrent physicalist argument, emphasized particularly by two-aspect theories, but used by other varieties of physicalism, is that an entity characterized by one set of properties may be identical with an entity characterized by an entirely different set of properties. The characterizations may involve different sensory modalities, or one may be sensory and the other microphysical. The argument then proceeds by suggesting that the identity of a mental state with a brain state (or with a class of brain states or with a program) is an instance of this general logic of identity. There seems to me to be a profound error in this reasoning. When an object characterized by one sensory modality is identified with one characterized by another modality, there is a tacit reference to two causal chains, both having a common terminus in a single object and another common terminus in the theatre of consciousness of the perceiver, but with different intermediate causal links in the environment and in the sensory and cognitive apparatus of the perceiver. When a brain state and a state of consciousness are identified, according to the two-aspect version of physicalism, there is no difficulty in recognizing a common object as terminus: it is, in fact, the brain state, since physicalism is committed to the ontological primacy of the physical description. But the other terminus, the theatre of consciousness of the perceiver, is absent. Or perhaps one should say that there is a pervasive equivocation in the two-aspect theory, since a common theatre is tacitly assumed as the locus of combination and comparison of the physical and the mental aspect, but on the other hand, there is no independent status for this theatre, if physicalism is correct.

A related argument against physicalism relies upon a philosophical principle which I call 'the phenomenological principle' (but would welcome a better name, if it exists in the literature or might be suggested): that is, whatever ontology a coherent philosophy recognizes, that ontology must suffice to account for appearances. This principle has the consequence that physicalism is incoherent. A physicalist ontology may, and usually does, postulate an ontological hierarchy, the fundamental level typically consisting of elementary particles or fields, the higher levels consisting of composites formed from the elementary entities. These composites may be characterized in different ways: fine-grained characterizations give the microstate in detail; coarse-grained characterizations sum or average or integrate over the fine-grained descriptions; relational characterizations depend on causal connections between the composite systems of interest and instruments or perceivers. Where do sensory appearances fit in this conception of nature? They do not fit in the fine-grained characterizations, unless mental properties are smuggled into fundamental physics, contrary to the program of physicalism. They do not fit in the coarse-grained description without something like the two-aspect theory, whose weakness was pointed out in the preceding paragraph; and they do not fit in the relational characterizations unless the object is causally connected to a sensitive subject. In sum, sensory appearances fit nowhere in a physicalist ontology.

These two arguments against physicalism are simple-minded but robust. It is hard to see how they could be resisted and how

147

mind could be viewed as ontologically derivative, were it not for several massive and formidable considerations. The first is that there is no evidence at all for mentality existing apart from highly developed nervous systems. As Roger says, 'If the "mind" is something quite external to the physical body it is hard to see why so many of its attributes can be very closely associated with properties of a physical brain' (SM p. 350). The second is the immense body of evidence that neural structures are products of evolution from primitive organisms devoid of such structures, and indeed, if the program of prebiotic evolution is correct, the genealogy can be extended back to inorganic molecules and atoms. The third consideration is that fundamental physics attributes no mental properties to these inorganic constituents.

A. N. Whitehead's 'philosophy of organism'[5] (which had Leibniz's monadology as an antecedent) has a mentalistic ontology that takes all of the three preceding considerations into account, but with subtle qualifications. Its ultimate entities are 'actual occasions', which are not enduring entities but spatio-temporal quanta, each endowed – usually on a very low level – with mentalistic characteristics like 'experience', 'subjective immediacy', and 'appetition'. The meanings of these concepts are derived from the high-level mentality that we know introspectively, but immensely extrapolated from this familiar base. A physical elementary particle, which Whitehead conceives as a temporal chain of occasions, can be characterized with very little loss by the concepts of ordinary physics, because its experience is dim, monotonous and repetitious; but nevertheless there is some loss: 'The notion of physical energy, which is at the base of physics, must then be conceived as an abstraction from the complex energy, emotional and purposeful, inherent in the subjective form of the final synthesis in which each occasion completes itself'.[6] Only the evolution of highly organized societies of occasions permits primitive

148

mentality to become intense, coherent and fully conscious: 'the functionings of inorganic matter remain intact amid the functionings of living matter. It seems that, in bodies that are obviously living, a coordination has been achieved that raises into prominence some functions inherent in the ultimate occasions'.[7]

Whitehead's name is not in the index of SM and its only occurrence in *The Emperor's New Mind*[8] refers to *Principia Mathematica* of Whitehead and Russell. I don't know the reasons for Roger's neglect of him, but I can state some objections of my own with which he might concur. Whitehead offers his mentalistic ontology as a remedy for the 'bifurcation of nature' into the mindless world of physics and the mind of high-level consciousness. The low level of protomentality that he attributes to all occasions is intended to bridge this enormous gap. But is there not a comparable bifurcation between the protomentality of elementary particles and the high-level experience of human beings? And is there any direct evidence at all for low-level protomentality? Would any one have postulated it except in order to establish continuity between the early Universe and the present Universe inhabited by conscious organisms? And if there is no reason other than this, would not the morpheme 'mental' in the word 'protomental' be an equivocation, and does not the entire philosophy of organism become a semantical trick of taking a problem and renaming it a solution? Furthermore, does not the conception of the actual occasions as the ultimate concrete entities of the Universe constitute a kind of atomism, richer to be sure than that of Democritus and Gassendi, but nevertheless inconsistent with the holistic character of mind that our high-level experience reveals?

In the following section I suggest that these objections can be answered to some extent by working out a modernised Whiteheadianism, using some concepts drawn from quantum mechanics.[9]

4.2 THE RELEVANCE OF QUANTUM THEORETICAL
IDEAS TO THE MIND-BODY PROBLEM

The most radical concept of quantum theory is that a complete state of a system – that is, one which specifies the system maximally – is not exhausted by a catalogue of actual properties of the system but must include potentialities. The idea of potentiality is implicit in the superposition principle. If a property A of a quantum system and a state vector ϕ (assumed for convenience to have norm unity) are specified, then ϕ can be expressed in the form $\sum_i c_i u_i$, where each u_i is a state vector of unit norm representing a state in which A has a definite value a_i, and where each c_i is a complex number with the sum of $|c_i|^2$ being unity. Then ϕ is a superposition of the u_i with appropriate weights, and unless the sum contains only a single term the value of A in the state represented by ϕ is indefinite. If the quantum state is interpreted realistically, as a representation of the system as it is, rather than as a compendium of knowledge about it, and if the quantum description is complete, not susceptible to any supplementation by 'hidden variables', then this indefiniteness is objective. Furthermore, if the system interacts with its environment in such a way that A becomes definite, for example, by means of measurement, then the outcome is a matter of objective chance, and the probabilities $|c_i|^2$ of the various possible outcomes are objective probabilities. These features of objective indefiniteness, objective chance and objective probability are summed up by characterizing the quantum state as a network of potentialities.

The second radical concept of quantum theory is entanglement. If u_i are state vectors of unit norm representing states of System I, with some property A having distinct values in these states, and v_i are state vectors of System II, with a property B having distinct values in them, then there is a state vector $X = \sum_i c_i u_i v_i$

150

(the $|c_i|^2$ summing to unity) of the composite System I + II with peculiar features. Neither I nor II separately is in a pure quantum state. In particular, I is not a superposition of the u_i and II is not a superposition of the v_i, for such superpositions omit the way in which the u_i and the v_i are correlated. X is thus a kind of holistic state, called 'entangled'. Quantum theory thus has a mode of composition without an analogue in classical physics. If a process occurs whereby A becomes actualized, for example, by having the value a_i, then B will automatically be actualized also and will have the value b_i. Entanglement thus entails that the potentialities of I and II are actualized in tandem.

The modernized Whiteheadianism that I cryptically referred to at the end of Section 4.1 incorporates the concepts of potentiality and entanglement in essential ways. Potentiality is the instrument whereby the embarrassing bifurcation between dim protomentality and high-level consciousness can be bridged. Even a complex organism with a highly developed brain may become unconscious. The transition between consciousness and unconsciousness need not be interpreted as a change of ontological status, but as a change of state, and properties can pass from definiteness to indefiniteness and conversely. In the case of a simple system like an electron one can imagine nothing more than a transition from utter indefiniteness of experience to a minimal glimmering. But at this junction the second concept, entanglement, comes into play. For a many-body system in entangled states there is a much richer space of observable properties than for a single particle, and the spectra of the collective observables are commonly much broader than those of the component particles. The entanglement of elementary systems each with a very narrow range of mental attributes can, conceivably, generate a broad range, all the way from unconsciousness to high level consciousness.

How does this modernized Whiteheadianism compare with Roger's application of quantum ideas to the mind–body problem? In Chapter 7 of SM and in Chapters 2 and 3, Roger makes essential use of the two great ideas of potentiality and entanglement. Potentiality is invoked in his conjecture that 'quantum computations' are performed by a system of neurons, each branch of a superposition performing a calculation independent of those performed in the other branches (SM pp. 355–6). Entanglement (which Roger usually refers to as 'coherence') is invoked in several stages in order to account for the performance of these calculations: the microtubules in cell walls are supposed to play an organizing role in the functioning of neurons, and for this purpose an entangled state of a microtubule is postulated (SM pp. 364–5); the microtubules of a single neuron are then supposed to be in an entangled state; and finally there is a putative entangled state of a large number of neurons. Large-scale entanglement is needed because 'the unity of a single mind can arise in this description only if there is some form of quantum coherence extending across an appreciable part of the entire brain' (SM p. 372). Roger maintains that his proposal is plausible in view of the phenomena of superconductivity and superfluidity, especially of high-temperature superconductivity, and of Fröhlich's calculations that large-scale entanglement is possible in biological systems at body temperature (SM pp. 367–8). One more quantum idea in Roger's treatment of mind is drawn not from current quantum theory but from the quantum theory of the future that he envisages, which will be discussed in Section 4.3. This idea is the objective reduction of a superposition (abbreviation **OR**), whereby an actual value of an observable A is selected from an initially broad range of possible values. That such actualization is indispensable for a theory of the mind is entailed by the undoubted phenomena of definite sensations and thought in our conscious experience. It is needed even if there is such a thing as quantum computation, because at the end of the parallel processing in the various branches of superposition a definite 'result' must be read out (SM p. 356). Finally, Roger conjectures that **OR** will supply the non-computational aspects of mental activity.

From a modernized Whiteheadian standpoint, what is missing – inadvertently or deliberately – in Roger's theory of mind is the idea of mentality as something ontologically fundamental in the Universe. Roger's account sounds suspiciously like a quantum version of physicalism. In the versions of physicalism referred to in Section 4.1, mental properties were treated as structural properties of brain states or as programs for performing calculations by neural assemblies. Roger supplies new ingredients for the program of physically accounting for mentality – namely, quantum coherence on a large scale and a putative modification of quantum dynamics in order to account for the reduction of superpositions. But this sophistication does not weaken the naive but robust arguments against physicalism offered in Section 4.1. The appearances of our mental life have no place in a physicalist ontology, and a physicalism governed by quantum rules is still physicalistic. Whitehead's philosophy of organism, by contrast is radically non-physicalistic, since it attributes mentalistic properties to the most primitive entities in the Universe, thus conjecturally enriching the physical description of them. The modernized version of Whiteheadianism which I tentatively proposed does not use quantum theory as a surrogate for the fundamental ontological status of mentality, but as an intellectual instrument for accounting for the immense gamut of manifestations of mentality in the world, from complete depression of the intrinsic mentality to high-level enhancement of it.

The contrast can be put another way. Quantum theory is a framework, deploying such concepts as state, observable, super-

153

position, transition probability and entanglement. Physicists have applied this framework successfully to two very different ontologies – the ontology of particles, in the standard non-relativistic quantum mechanics of electrons, atoms, molecules and crystals; and the ontology of fields, in quantum electrodynamics, quantum chromodynamics and general quantum field theory. Conceivably quantum theory can be applied to entirely different ontologies, such as an ontology of minds, a dualistic ontology, or an ontology of entities endowed with protomentality. The usual physicalistic applications of quantum theory have been wonderfully fertile in accounting for observable phenomena of composite systems, including macroscopic ones, in microphysical terms. It seems to me that Roger is trying to do something similar, by explaining mental phenomena in a physicalistic ontology through delicate employment of quantum concepts. Modernized Whiteheadianism, by contrast, applies the framework of quantum theory to an ontology that is *ab initio* mentalistic. Admittedly, modernized Whiteheadianism is inchoate, impressionistic and devoid of clean theoretical predictions and experimental confirmations that would establish its credentials as a 'promising' theory. But it has the great virtue of recognizing the underivability of mentality, which is missing in all varieties of physicalism. It may be that I have misread or misheard Roger, and that in fact he is more of a crypto-Whiteheadian than I had realized. Whether this is so or not, his explicitness on the matter would greatly clarify his position.

If a modernized version of Whitehead, or any quantum theory of mind whatever, is to achieve scientific maturity and solidarity, then much attention will have to be paid to psychological phenomena. There are some phenomena that have a 'quantum flavor' about them: for example, transitions from peripheral to focal vision; transitions from consciousness to unconsciousness;

the pervasiveness of the mind through the body; intentionality; anomalies in temporally locating mental events; and the conflations and ambiguities of Freudian symbolism. Several important books on the relation between quantum theory and mind have examined mental phenomena that have a quantum flavor, notably those of Lockwood[10] and Stapp.[11] Roger himself discusses some of these phenomena, for instance Kornhuber's and Libet's experiments on timing passive and active aspects of consciousness (SM pp. 385–7).

A serious application of quantum theory to the mind must also consider the mathematical structure of the space of states and the set of observables. These are not supplied by the quantum framework. In the case of standard non-relativistic quantum mechanics and quantum field theory these structures are determined in various ways: by considerations of the representation of space-time groups, by heuristics based on classical mechanics and classical field theory and of course by experiment. One of Schrödinger's great papers on wave mechanics of 1926 presents a wonderfully fruitful analogy: geometrical optics is to wave optics as particle mechanics is to a hypothetical wave mechanics. Might it not be heuristically valuable to consider a new analogy: classical physics is to quantum physics as classical psychology is to a hypothetical quantum psychology? Of course, one of the difficulties of exploiting this analogy is that the structure of 'classical psychology' is much less well known and perhaps less definite intrinsically than the structure of classical mechanics.

Here is one further suggestion. Possibly quantum concepts can be applied to psychology, but not with as much geometrical structure as in quantum physics. Even if there is such a thing as a space of mental states, can we assume that this space will have the structure of a projective Hilbert space? In particular, will an inner product be defined between any two mental states, which will determine the transition probability from one to the other? Might it not be the case that a weaker structure exists in nature, though a structure of quantum kind? There are very interesting papers by Mielnik[12] suggesting that a minimal quantum concept is the expressibility of a 'mixed' state in more than one way as a convex combination of pure states, whereas in classical statistical mechanics a mixed state can be expressed in only one way in terms of pure states. A further speculation is that the phenomenology of colours can be constructed as exemplifying Mielnik's idea – for example, many different ways of composing perceptual white out of a mixture of coloured light.

4.3 THE PROBLEM OF THE ACTUALIZATION OF POTENTIALITIES

In Chapter 2 Roger classified the problem of the actualization of potentialities (also called the problem of the reduction of the wave packet and the measurement problem) as an X-mystery, one that cannot be solved without a radical change of the theory itself rather than one which can be exorcized by habituation. I fully agree. If quantum theory objectively describes a physical system, then there are observables of the system that are objectively indefinite in a specified state but which become definite when a measurement is performed. But the linear dynamics of quantum theory precludes actualization by means of measurement. Linearity has the consequence that the final state of the composite system of measuring apparatus plus object is a superposition of terms in which the 'pointer' observable of the apparatus has different values. I share Roger's scepticism about all attempts to interpret away this mystery, for example, by many-worlds interpretations, decoherence, hidden variables, etc. At some stage or another in a measurement process the unitary evolution of the quantum state breaks down and an actualization occurs. But at what stage? There are many possibilities.

The stage may be physical, and it may occur when a macroscopic system is entangled with a microscopic object, or when the space-time metric is entangled with a material system. Or the stage may be mental, occurring in the psyche of the observer. Roger hypothesizes that actualization is a physical process, due to the instability of a superposition of two or more states of the space-time metric; the greater is the energy difference among the superposed states, the shorter is the lifetime of the superposition (SM pp. 339–46). However, the conjunction of this conjecture with Roger's determination to account for actual experiences in consciousness imposes some strenuous constraints. He needs the superposition of brain states, as indicated earlier, to account for the globality of mind, but such monstrosities as the superposition of seeing a red flash and seeing a green flash must either not occur at all or be so transient as hardly to impinge upon consciousness. Roger argues – tentatively and sketchily – that the energy differences in the brain states corresponding to such distinct perceptions are sufficiently great to yield a short lifetime of the superposition. However, he admits in a number of passages (SM pp. 409, 410, 419, 342–3) that he is trying to perform a delicate tight-rope walk, for he must maintain enough coherence to account for the globality of mind and enough breaking of coherence to account for definite conscious events. How a brain/mind operating along the lines sketched by Roger could be robust in daily operation is very mysterious indeed.

The resources of the family of modifications of quantum dynamics for the purpose of accounting objectively for the actualization of potentialities have not yet been fully explored, either by Roger or by the community of investigators. I shall mention

briefly two avenues which I find attractive. The spontaneous reduction model of Ghirardi–Rimini–Weber and others is mentioned
by Roger and cogently criticized (SM p. 344), but there may be vari-
ants of this dynamics that will escape his criticisms. A second
avenue, which he does not mention, is the possibility of a 'superselection rule' in nature, preventing the superposition of distinct
isomers or conformations of macromolecules. The motivation for
this conjecture is the consideration that macromolecules typically
act as switches in the cell, turning off or on processes according to
molecular conformation. If two distinct conformations were superposed, one would have a cellular analogue of Schrödinger's cat
– a process in limbo between occurring and not occurring. If nature obeys a superselection rule prohibiting such superpositions,
embarrassment would be avoided, but the reason would be mysterious: why does nature prohibit superpositions of conformation
states of complex molecules when it allows them for simple ones,
and where is the dividing line? However, such a superselection
might account for all the actualizations of potentialities for which
we have good evidence, and it may have the precious property of
being testable by molecular spectroscopy.[13]

Finally, it is worth remarking that from a Whiteheadian point
of view the hypothesis that the actualization of potentialities is
achieved by the psyche of the perceiver is not as ridiculous, an-
thropocentric, mystical and unscientific as it is commonly regarded to be. According to Whitehead, something like mentality
is pervasive throughout nature, but high-level mentality is contingent upon the evolution of special hospitable complexes of
occasions. The capacity for a system to actualize potentialities,
thereby modifying the linear dynamics of quantum mechanics,
may be pervasive in nature, but non-negligible only in systems
with high-level mentality. I would qualify this expression of tol-
eration, however, by saying that the attribution of the power of

reducing superpositions to the psyche should be taken seriously only if its implications for a wide range of psychological phenomena are carefully worked out, for only then would there be a possibility of subjecting the hypothesis to controlled experimental test.

NOTES

1. 'We have to know, so we will know'. This exhortation is engraved on Hilbert's gravestone. See Constance Reid (1970). *Hilbert*, p. 220. (New York: Springer-Verlag).

2. Hilary Putnam (1994) Review of *Shadows of the Mind, The New York Times Book Review*, Nov. 20 1994, p. 1.

3. Roger Penrose (1994) Letter to *The New York Times Book Review*, Dec. 18 1994, p. 39.

4. Ned Block (1980) *Readings in Philosophy of Psychology*, Volume 1, Parts 2 and 3. (Harvard University Press, Cambridge, MA).

5. Alfred North Whitehead (1933) *Adventures of Ideas*, (Macmillan, London) (1929) *Process of Reality* (Macmillan, London).

6. A. N. Whitehead, *Adventures of Ideas*, Chapter 11, Section 17.

7. Ibid., Chapter 13, Section 6.

8. Roger Penrose (1989) *The Emperor's New Mind*. (Oxford University Press, Oxford).

9. Abner Shimony (1965) 'Quantum physics and the philosophy of Whitehead', in Max Black (ed.), *Philosophy in America* (George Allen & Unwin, London): reprinted in A. Shimony (1993). *Search for a Naturalistic World View*, Volume 2, pp. 291–309. (Cambridge University Press, Cambridge); Shimon Malin, (1988). A Whiteheadian approach to Bell's correlations, *Foundations of Physics*, **18**, 1035.

10. M. Lockwood (1989) *Mind, Brain and the Quantum,* (Blackwell, Oxford).

11. Henry P. Stapp (1993) *Mind, Matter and Quantum Mechanics* (Springer-Verlag, Berlin).

12. Bogdan Mielnik (1974) Generalized quantum mechanics, *Communications in Mathematical Physics,* **37**, 221.

13. Martin Quack (1989) Structure and dynamics of chiral molecules, *Angew. Chem. Int. Ed. Engl.* **28**, 571.

CHAPTER 5

Why Physics?

NANCY CARTWRIGHT

We discussed Roger Penrose's book *Shadows of the Mind* in a joint LSE/King's College London seminar series *Philosophy: Science or Theology*. I want to begin by asking the same question that was asked of me by one of the participants in the seminar – 'What are Roger's reasons for thinking answers to questions about the mind and consciousness are to be found in physics rather than in biology?' So far as I could see there are three kinds of reasons that Roger suggests:

(1) We can lay out a very promising programme for doing it that way. This is potentially the most powerful kind of reason one can give for a project like Roger's. Indeed positivist that I am, opposed at once both to metaphysics and to transcendental argument, I would be prepared to argue that it is the only kind of argument to which we should give much weight. Of course, how strongly this kind of argument supports a project will depend on how promising the programme is – and how detailed. One thing that is clear is that Roger's proposal – first to posit macroscopic quantum coherence across the microtubules of the cytoskeleton and then to look for the special non-computational features of consciousness in a new kind of quantum–classical interac-

tion – is not a detailed programme. Its promise certainly does not lie in the fact that it is a natural next step in a well-verified progressive research agenda. If one finds it promising, it must be due to the boldness and imaginativeness of the ideas, to the conviction that some new interaction of this kind is necessary anyway to sort out quantum mechanics, and to the strong prior commitment that, if there is to be a scientific explanation for consciousness, it must ultimately be a *physics* explanation. I think this last must surely play a key role if we are to judge Roger's programme promising. But obviously, to the extent that it does play a role, the fact that we do judge the programme promising cannot supply us with a reason for thinking it is physics, and not some other science, that will do the job.

(2) The second kind of reason for thinking physics by itself will provide the ultimate account is the undoubted fact that bits of physics – especially electromagnetism – contribute to our understanding of the brain and of the nervous system. By now we standardly describe message transmission using concepts of electrical circuitry. Part of Roger's own story relies on quite recent imports of electromagnetism: different states of electrical polarization in a tubulin dimer are supposed to be the basis for differences in geometrical configuration that cause the dimers to bend at different angles to the microtube. But this kind of argument won't do. The fact that physics tells part of the story is poor reason for concluding it must tell the whole story.

Sometimes chemistry is brought up at this stage to argue the contrary. Now no-one would deny that a chunk of the story will be told by chemistry. But the relevant bits of chemistry are themselves really just physics, it is supposed. This is very much the way Roger himself talks about it: 'The chemical forces that control the interaction of atoms and molecules are indeed quantum mechanical in origin, and it is largely chemical action that governs the behaviour of the *neurotransmitter* substances that transfer signals from one neuron to another – across tiny gaps that are called *synaptic clefts*. Likewise, the action potentials that physically control nerve-signal transmission itself have an admittedly quantum mechanical origin' (SM p. 348). Chemistry came into play in defence of physics in response to my worries about the gigantic inferential leap from 'Physics tells part of the story' to 'Physics tells all of the story'. But now this same inferential leap has reappeared all over again one level down. Notoriously we have nothing like a real reduction of the relevant bits of physical chemistry to physics – whether quantum or classical.[1] Quantum mechanics is important for explaining aspects of chemical phenomena but always quantum concepts are used alongside of *sui generis* – that is, unreduced – concepts from other fields. They don't explain the phenomena on their own.

(3) The third reason for thinking physics will explain the mind is metaphysical. We can see Roger's chain of connection. We should like to assume that the function of the mind is *not mysterious*; that means it can be explained in *scientific terms*; that means it can be explained in *physics* terms. In my seminar the question, 'Why not biology?' was raised by the well-known statistician James Durbin. That I think is relevant. As a statistician, Durbin lives in a mottled world. He studies patterns of characteristics that come from all sorts of fields, both scientific and practical. Roger's world by contrast is the world of the *unified system*, with physics as the base for unification. The reason, I think, for this kind of physics-ism is the idea that we have no satisfactory

metaphysics otherwise. Without the system we are left with some kind of unacceptable, or to use Roger's word – mysterious – dualism. That is the topic that I want to discuss,[2] for I think the view that there is no sensible alternative is one that has a real grip on many physicists. There is the feeling that anyone who takes physics seriously as really describing the world will have to believe in its hegemony.

Why? There are apparently a very, very large number of different properties at work in the world. Some are studied by one scientific discipline, some by another, some are in the intersection of different sciences, and most are not studied by any science at all. What legitimates the view that behind the appearances they are all really the same? I think two things: one is an excessive confidence in the systematicity of their interactions, and the other, an excessive estimation of what physics has accomplished.

I should remark though that this limitation in metaphysical vision that sees as possible only a kind of physics-istic monism is also widespread in philosophy, even among those who resist the reduction of the special sciences to physics. Consider philosophy of biology where reductionism has long been out of fashion and now again a kind of emergenticism is taken seriously, with properties and laws arising newborn with increasing levels of complexity and organization. Still most cannot get beyond a kind of monism; they feel compelled to insist on 'supervenience'. Roughly, to say the properties of biology supervene on those of physics is to say that if we had two situations that were identical with respect to their physical properties they must be identical with respect to their biological properties. This doesn't mean, they say, that biological laws are reduced to physics laws since biological properties needn't be definable in physics terms. But it does mean that biological properties are not separate independent properties on their own, for they are fixed by the properties of physics. Once the physics description is set, then the biological description cannot be but what it is. The biological properties do not have full independent status. They are second class citizens.

To take seriously that biological properties are separate properties, causally effective on their own, is not to fly in the face of the empirical evidence. I take for granted what we see in science: sometimes physics helps explain what goes on in biological systems. But as I said of chemistry, the same here: seldom without the aid of unreduced, *sui generis* biological descriptions as well. We can turn a slogan I have used in a different form elsewhere: no biology in, not biology out.† What we see is most naturally

† During the discussion, Abner Shimony made the following remarks
in relation to this issue:
'Nancy Cartwright argued for discussing mind in the context of biology rather than of physics. I applaud the positive part of her request. Of course, there is much to be learned about the mind from evolutionary biology, anatomy, neurophysiology, developmental biology, etc. But I do not agree that the investigation of the relation of mind to physics is sterile. Connections among disciplines should be pursued as deeply as possible; relations between wholes and parts should be pursued as deeply as possible. One does not know *a priori* where these investigations will lead, and in different domains the results have been very different. Thus, Bell's Theorem and the experiments which it inspired have shown that correlations exhibited by spatially separated entangled systems cannot be accounted for by any theory that ascribes definite states to the individual system – a great triumph for holism. The proof by Onsager that the two-dimensional Ising model undergoes phase transitions shows that long-range order can be exhibited in an infinite system in which the components only interact with their nearest neighbours – a triumph for the analytic point of view and for the reducibility of macrophysics to microphysics. Either type of discovery – holistic or analytic – reveals something important about the world. The investigation of relations among disciplines does not infringe the validity of phenomenological laws within disciplines. Such investigations may provide heuristics for refined phenomenological laws, and also may offer a deepened understanding of such laws. When Pasteur suggested that the chirality of molecules is responsible for rotation of the plane of polarization of light passing through solutions, he founded stereo-chemistry'.

described as interaction between biological and physics characteristics, each affecting the other. We also have very context-local identifications of a biological and physical description, as well as a good deal of causal cooperation – biological and physical properties acting together to produce effects neither can cause on their own. To go from that to 'It must all be physics' is just the gigantic inferential leap I've been worrying about. What we see may be consistent with its all being physics but it certainly doesn't single out that conclusion, and indeed, on the face of it, it seems to point away from it.[3]

Part of the reason for thinking that it must all be physics is, I believe, a view about closure. The concepts and laws of a good physics theory are supposed to constitute a system closed on itself: it is all you need to be able to make predictions about these very concepts themselves. I think this is a mistaken – or at least unwarrantedly optimistic – view of the success of physics. About the same time that the idea of supervenience became prominent in philosophy, so too did the idea of a special science. Essentially all sciences except physics are special sciences. That means that their laws hold at best only *ceteris paribus*. They hold only so long as nothing from outside the domain of the theory in question interferes.

But what generates the confidence that the laws of physics are more than *ceteris paribus* laws? Our amazing laboratory successes show no such thing; nor does the Newtonian success with the planetary system, which so impressed Kant. And neither do the great technical exports from physics – vacuum tubes or transistors or SQUID magnetometers. For these devices are built to ensure that no interference occurs. They don't test whether the laws are still good when factors from outside the domain of the theory play a role. There is of course the general faith that, in the case of physics, nothing could interfere except further factors that themselves can be described in the language of physics and that are subject to its laws. But that of course is just the point at issue.

I want to close with a remark about realism. I've been pointing to a kind of pluralistic view of all the sciences standing side by side on a roughly equal footing with various different kinds of interactions between the factors studied in their different domains. This is a picture that often goes along with a view that science is a human construction that does not mirror nature. But this is not a necessary connection. Kant had exactly the opposite stance: it is precisely because we construct science that the unified system is not only possible but necessary. Nevertheless nowadays this pluralistic picture is often associated with social constructionism. So it is important to stress that pluralism does not imply anti-realism. To say the laws of physics are true *ceteris paribus* is not to deny that they are true. They are just not entirely sovereign. It is not realism about physics that is at stake under pluralism, but rather imperialism. So I want not to point us to a discussion of scientific realism. Rather I want Roger to discuss his metaphysical commitment that it must be physics that will do the job. For that must be presupposed if the discussion is already about whether it will be this kind of physics or that. The issue isn't whether the laws of physics are true and do in some ways bear on the operation of the mind, but whether they are all that's true or must carry the bulk of the explanatory burden.

NOTES

1. See R. F. Hendry: Approximations in quantum chemistry in Niall Shanks (ed.), *Idealisation in Contemporary Physics*, (Poznań Studies in the Philosophy of the Sciences and Humanities, Rodopi, Amsterdam) (forthcoming 1997).

R.G. Woolley (1976): 'Quantum theory and molecular structure', *Advances in Physics*, **25**, 27-52.

2. For details of arguments against the single system, see John Dupre (1993) *The Disorder of Things: Metaphysical Foundations of Disunity of Science* (Harvard University Press, Cambridge MA); Otto Neurath (1987) *Unified Science*, Vienna Circle Monograph Series, trans. H. Kael (D. Reidel: Dordrecht).

3. For a further discussion of this point, see Nancy Cartwright (1993) Is natural science natural enough? A reply to Phillip Allport, *Synthese*, **94**, 291. For a more elaborated discussion of the general point of view mooted here, see Nancy Cartwright (1994) 'Fundamentalism vs the patchwork of laws', *Proceedings of the Aristotelian Society* and (1995) 'Where in the world is the quantum measurement problem', *Physik, Philosophie und die Einheit der Wissenschaft, Philosophia Naturalis*, ed. L. Kreuger and B. Falkenburg (Spektrum: Heidelberg).

CHAPTER 6

The Objections of an
Unashamed Reductionist
STEPHEN HAWKING

To start with, I should say I'm an unashamed reductionist. I believe that the laws of biology can be reduced to those of chemistry. We have already seen this happening with the discovery of
the structure of DNA. And I further believe that the laws of chemistry can be reduced to those of physics. I think most chemists
would agree with that.

Roger Penrose and I worked together on the large-scale structure of space and time, including singularities and black holes. We
pretty much agree on the classical theory of General Relativity but
disagreements began to emerge when we got on to quantum gravity. We now have very different approaches to the world, physical and mental. Basically, he's a Platonist believing that there's
a unique world of ideas that describes a unique physical reality.
I, on the other hand, am a positivist who believes that physical
theories are just mathematical models we construct, and that it
is meaningless to ask if they correspond to reality, just whether
they predict observations.

This difference in approach has led Roger to make three claims
in Chapters 1-3 that I strongly disagree with. The first is that
quantum gravity causes what he calls **OR**, objective reduction of
the wavefunction. The second is that this process has an important role in the operation of the brain through its effect on coher-

ent flows through microtubules. And the third is that something like **OR** is needed to explain self-awareness because of the Gödel Theorem.

To start with quantum gravity, which is what I know best. His objective reduction of the wavefunction is a form of decoherence. This decoherence can come about through interactions with the environment or through fluctuations in the topology of space-time. But Roger seems to want neither of these mechanisms. Instead he claims that it occurs because of the slight warping of space-time produced by the mass of a small object. But, according to accepted ideas, that warping will not prevent a Hamiltonian evolution with no decoherence or objective reduction. It may be that accepted ideas are wrong but Roger has not put forward a detailed theory that would enable us to calculate when objective reduction would occur.

Roger's motivation in putting forward objective reduction seems to have been to rescue Schrödinger's poor cat from its half-alive, half-dead state. Certainly, in these animal liberation days, no one would dare suggest such a procedure, even as a thought experiment. However, Roger made a point of claiming that objective reduction was so weak an effect that it could not be experimentally distinguished from decoherence caused by interaction with the environment. If that is the case, then environmental decoherence can explain Schrödinger's cat. There's no need to invoke quantum gravity. Unless objective reduction is a strong enough effect to be measured experimentally, it can't do what Roger wants it to do.

Roger's second claim was that objective reduction had a significant influence on the brain, maybe through its effect on coherent flows through microtubules. I'm not an expert on the operation of the brain, but it seems very unlikely, even if I believed in objective reduction, which I don't. I cannot think that the brain contains systems that are sufficiently isolated that objective reduction could be distinguished from environmental decoherence. If they were that well isolated they wouldn't interact rapidly enough to take part in mental processes.

Roger's third claim is that objective reduction is somehow necessary because Gödel's Theorem implies that a conscious mind is not computable. In other words, Roger believes that consciousness is something special to living beings and that it couldn't be simulated on a computer. He didn't make it clear how objective reduction could account for consciousness. Rather, his argument seemed to be that consciousness is a mystery and quantum gravity is another mystery so they must be related.

Personally, I get uneasy when people, especially theoretical physicists, talk about consciousness. Consciousness is not a quality that one can measure from the outside. If a little green man were to appear on our door step tomorrow, we do not have a way of telling if he was conscious and self-aware or was just a robot. I prefer to talk about intelligence which is a quality that can be measured from the outside. I see no reason why intelligence should not be simulated on a computer. We certainly can't simulate human intelligence at the moment, as Roger showed with his chess problem. But Roger also admitted that there was no dividing line between human intelligence and animal intelligence. So it will be sufficient to consider the intelligence of an earthworm. I don't think there's any doubt that one can simulate an earthworm's brain on a computer. The Gödel argument is irrelevant because earthworms don't worry about Π_1-sentences.

The evolution from earthworm brains to human brains presumably took place by Darwinian natural selection. The quality selected for was the ability to escape enemies and to reproduce, not the ability to do mathematics. So again the Gödel Theorem is not relevant. It is just that the intelligence needed for survival can also be used to construct mathematical proofs. But it is a very

171

hit and miss business. We certainly don't have a knowably sound procedure.

I have told you why I disagree with Roger's three claims that there is objective reduction of the wavefunction, that this plays a role in the operation of the brain and that it is necessary to explain consciousness. I had better let Roger reply.

CHAPTER 7

Roger Penrose Responds

I am grateful for the comments by Abner, Nancy and Stephen, and I wish to make a few remarks in response. In what follows, I shall be replying separately to each of them.

RESPONSE TO ABNER SHIMONY

First, let me say that I very much appreciate Abner's comments, which I think are extremely helpful. However, he suggests that, by concentrating on the computability issue, I may be attempting to climb the wrong mountain! If, by this, he is pointing out that there are many important manifestations of mentality other than non-computability, then I fully agree with him. I agree, also, that Searle's Chinese room argument provides a convincing case against the 'strong-AI' position that computation alone can evoke conscious mentality. Searle's original argument was concerned with the mental quality of 'understanding', as is my own 'Gödelian' discussion, but the Chinese room can also be used (perhaps with even greater force) against other mental qualities, such as the sensation of a musical sound or the perception of the colour red. The reason that I have not used this line of argument in my own discussion, however, is that it is of an entirely negative character, and it does not give us any real clue as to what is actually going on

173

with consciousness, nor does it indicate any direction in which we should proceed if we are to attempt to move towards a scientific basis for mentality.

Searle's line of reasoning is concerned solely with the **A/B** distinction, in the terminology that I adopted in Chapter 3 (cf. also, *Shadows* pp. 12–16). That is to say, he wishes to show that the *internal* aspects of consciousness are not encapsulated by computation. This is not enough for me, because I need to show that the *external* manifestations of consciousness are not attainable by computation either. My strategy is not to attempt to tackle the much more difficult internal problems at this stage, but to try to do something more modest at first, by attempting to understand what kind of physics could conceivably give rise to the kind of external behaviour that can be exhibited by a conscious being – so it is the **A/C** or **B/C** distinction that concerns me at this stage. My case is that some progress is indeed possible here. Agreed, I am not yet attempting to mount a major assault on the *true* summit, but it is my belief that if we can first successfully negotiate one of its significant foothills, we shall then be much better able to see the way up to the actual summit from our new vantage point.

Abner refers to my letter(s) of response to Hilary Putnam's review of *Shadows*, remarking that he is not convinced by what I had to say. In fact, I made no real attempt to answer Putnam in detail, because I did not believe that letters pages in a magazine were the appropriate place to enter into a detailed discussion. I just wanted to point out that, in my opinion, Putnam's criticisms were a travesty. They were particularly irritating because he gave no indication of having even read those portions of the book that were aimed at the very points that he raised. There will be a much more detailed response in the (electronic) journal *Psyche* addressing a number of different reviews of *Shadows*, which I hope answers the points that Abner is worried about.† In fact, I believe that the 'Gödelian' case is, at root, really a very powerful one, even though some people seem most reluctant to take it on board. I am not going to give up on what I believe to be a basically correct argument merely because of certain people's difficulties with it! My point is that it supplies us with an important clue as to what kind of physics could possibly underlie the phenomenon of consciousness, even though this alone will certainly not tell us the answer.

I think that I am basically in agreement with the positive points that Abner makes. He is puzzled as to the lack of mention of A. N. Whitehead's philosophical work in either *Emperor* or *Shadows*. The main reason for this is ignorance, on my part. I do not mean to say that I was unaware of Whitehead's general position, whereby he holds to a form of 'panpsychism'. I mean that I had not read any of Whitehead's philosophical work in any detail and so would have been reluctant to comment on it or on its closeness or otherwise to my own thinking. I think that my general position is not out of line with what Abner is setting forth, though I had not been prepared to make any definite statement along these lines, partly owing to a lack of clear conviction as to what I actually do believe.

I find Abner's 'modernized Whiteheadianism' particularly striking, with a suggestive plausibility about it. I realize, now, that the kind of thing that must have been at the back of my mind is very close to what Abner so eloquently expresses. Moreover, he is right that large-scale *entanglements* are necessary for the unity of a single mind to arise as some form of collective quantum state. Although I had not explicitly asserted, in either *Emperor* or *Shad-*

† [Now appeared: January 1996; http://psyche.cs.monash.edu.au/ psyche-index-v2_1.html, and there is now a printed version, published by MIT press (1996)].

ows, the need for mentality to be 'ontologically fundamental in the Universe', I think that something of this nature is indeed necessary. No doubt there is some kind of protomentality associated with every occurrence of **OR**, according to my own view, but it would have to be exceedingly 'tiny' in some appropriate sense. Without some widespread entanglement with some highly organized structure, superbly adapted to some kind of 'information processing capability' – as occurs in brains – genuine mentality would presumably not significantly arise. I think that it is only because my own ideas are so ill formulated here that I did not venture any clearer statements as to my position on these matters. I am certainly grateful to Abner for his clarifying comments.

I also agree that there may be some significant insights to be gained from exploring possible analogies and experimental findings from the subject of psychology. If quantum effects are indeed fundamental to our conscious thought processes, then we ought to begin to see some of the implications of this fact in aspects of our thinking. On the other hand, one must be exceedingly careful, in this kind of discussion, not to jump to conclusions and pick up on false analogies. The whole area is a hotbed, and full of potential traps, I am sure. It may be that there are reasonably clear-cut experiments that could be performed, however, and it would be interesting to explore such possibilities. Of course, there might well be other types of experimental tests that one could perform that would be more specific to the microtubule hypothesis.

Abner mentions Mielnik's non-Hilbertian quantum mechanics. This kind of generalization of the framework of quantum theory has always struck me as interesting, and it is something that I believe should be studied further. I am not altogether convinced that it is precisely the sort of generalization which is needed, however. Two aspects of this particular idea make me uneasy. One of these is that, as with a number of other approaches to (generalizing) quantum mechanics, it concentrates, in effect, on the *density matrix*, rather than on the quantum state, as the way in which to describe reality. In ordinary quantum mechanics, the space of density matrices constitutes a convex set, and the 'pure states', which would have a single state-vector description, occur on the boundary of this set. This picture arises from an ordinary Hilbert space, being a subset of the tensor product of the Hilbert space and its complex conjugate (i.e., dual). In Mielnik's generalization, this general 'density matrix' picture is retained, but there is no underlying linear Hilbert space from which the convex set is constructed. I like the idea of generalizing away from the notion of a linear Hilbert space, but I am uneasy about losing the holomorphic (complex-analytic) aspects of quantum theory, which loss seems to be a feature of this approach. One does not retain an analogue of a state vector, as far as I can make out, but only of a state vector up to phase. This makes the complex superpositions of quantum theory particularly obscure within the formalism. Of course, it could be argued that it is these superpositions that cause all the trouble at the macroscopic scale, and maybe one should get rid of them. Nevertheless, they are quite fundamental at the quantum level, and I think that in this particular way of generalizing things we may be losing the most important positive part of quantum theory.

My other source of unease has to do with the fact that the nonlinear aspects of our generalized quantum mechanics ought to be set up to deal with the measurement process, there being an element of *time-asymmetry* involved here (see *Emperor*, Chapter 7). I do not see this aspect of things playing a role in Mielnik's scheme as it stands.

Finally, I should like to express my support for the quest for better theoretical schemes in which the basic rules of quantum mechanics are modified, and also for experiments that might be able to distinguish such schemes from conventional quantum theory. So far, I have not come across any suggestion for a presently feasible experiment that would be able to test the specific type of scheme that I promoted in Chapter 2. We are still a few orders of magnitude short, so far, but perhaps someone will come up with a better idea for a test.

RESPONSE TO NANCY CARTWRIGHT

I am encouraged (and flattered) to hear that *Shadows* has been discussed in a serious way in the LSE/King's College series to which Nancy refers. However, she voices a scepticism that one should try to answer questions concerning the mind in terms of physics rather than biology. I should first make clear that I am certainly not saying that biology is unimportant in our attempts to address this question. In fact, I think it is likely that really significant advances, in the near future, are more likely to be made on the biological than on the physical side – but mainly because what we need from physics, in my opinion, is a major revolution; and who knows when that will come!

But I suppose that this kind of concession is not what she means – as something that would count towards my regarding biology as capable of providing 'the fundamental ingredient' of the understanding of mentality in scientific terms. Indeed, from my own standpoint, it might be possible to have a conscious entity that is not biological at all, in the sense that we use the term 'biology' at the present time; but it would not be possible for an entity be conscious if it did not incorporate the particular type of *physical* process that I maintain is an essential.

Having said this, I am not at all clear as to what Nancy's position is with regard to the kind of line that is to be drawn between biology and physics. I get the view that she is being rather pragmatic about these issues, saying that it's OK to regard consciousness as a physics problem if this helps us to make progress. So she asks: can I really point to a specific research programme where physicists, rather than biologists, can help us to move forward in fundamental ways? I think that my proposals do lead to a much more specific programme than she seems to be suggesting. I claim that we must search for structures in the brain with some very clearcut physical properties. They should be such as to enable well-shielded spatially extended quantum states to exist, persisting for at least something of the general order of a second, where the entanglements involved in this state give it a spread over fairly large areas of the brain, probably involving many thousands of neurons all at once. To support such a state, we need biological structures with very precise internal construction, probably with a crystal-like structure, and able to have an important influence on synapse strengths. I do not see that ordinary nerve transmission can be sufficient on its own because there is no real chance of obtaining the needed isolation. Things like presynaptic vesicular grids, as has been suggested by Beck and Eccles, could be playing a role, but to my way of thinking, cytoskeletal microtubules appear to have more of the relevant qualities. It may be that there are many other structures on this sort of scale (such as clathrins) which are needed for the full picture. Nancy is suggesting that my picture is not very detailed; but it seems to me that it is a lot more detailed than almost any other that I have seen, and it has the potential to be worked out further in a very specific way, with many opportunities for experimental testing. I agree that there is much that is needed before we can get close to a 'complete' picture – but I think that we must move forward cautiously, and I do not expect definitive tests for a while yet. This is something that needs more work.

Nancy's more serious point seems to have more to do with the role that she sees physics playing in our overall world-view. I think, perhaps, that she views the status of physics as being overrated. Perhaps it is overrated – or at least the world-view that present-day physicists tend to present may well be grossly overstated as to its closeness to completion, or even as to its correctness!

Seeing (validly, in my view) that present-day physical theory is a patchwork of theories, Nancy suggests that it may always remain so. Perhaps the physicist's ultimate goal of a completely unified picture is an indeed unattainable dream. She takes the view that it is metaphysics, not science, even to address such a question. I am not sure, myself, what attitude to take on this, but I don't think that we really need to go that far in considering what is needed here. Unification has been a manifest overall trend in physics, and I see every reason to expect that this trend will continue. It would require a bold expression of scepticism to assert otherwise. Let us take what I consider to be the major piece of 'patchwork' in modern physical theory, namely the way in which the classical and quantum levels of description are stitched together – very unconvincingly, in my view. One could take the line that we must simply learn to live with two basically incompatible theories which apply at two different levels (which, I suppose, was more or less Bohr's expressed view). Now, we may be able to get away with such an attitude for some years to come, but as measurements get more precise and begin to probe the borderline between these two levels, we shall be wanting to know how Nature actually deals with this borderline. Perhaps the way in which some biological systems behave might depend critically on what goes on at this borderline. I suppose the question is whether we expect to find a beautiful mathematical theory to cope with what seems to us now to be an awkward mess, or is physics itself 'really' just an unpleasant mess at that level. Surely not! There is no question where my own instincts lie here on this question.

I get the impression from Nancy's remarks, however, that she might be prepared to accept just an unpleasant mess in the laws of physics at this stage.† Perhaps that's one thing that she might mean by biology not being reducible to physics. Of course, there may well be a great many complicated unknown parameters playing important roles at this level in biological systems. In order to handle such systems, even when all the underlying physical principles are known, it may be necessary in practice to adopt all sorts of guesses, approximation procedures, statistical methods and perhaps new mathematical ideas in order to provide a reasonably effective scientific treatment. But, from the standard physics point of view, even though the details of a biological system might provide us with an unpleasant mess, this is not a mess in the underlying physical laws themselves. If the physical laws

---

† During the discussion, Nancy Cartwright reiterated her position on this issue:
'Roger thinks a physics that cannot treat open systems is a bad physics. I, by contrast, think that it may well be a very good physics indeed – if the laws of nature are a patchwork, as I imagine they may be. If the world is full of properties that are not reducible to those of physics, but which causally interact with those that are, then the most accurate physics will necessarily be a *ceteris paribus* physics that can tell the whole story only about closed systems.

Which of these points of view is likely to be right? That, I take it, is a metaphysical question, metaphysical in the sense that any answer to it goes far beyond the empirical evidence we have, including that of the history of science. I urge avoiding this kind of metaphysics whenever possible, and, when methodological decisions require a commitment one way or the other, to hedge our bets heavily. Where we must bet, I would estimate the probabilities very differently from those who put their faith entirely in physics. Modern science is a patchwork not a unified system. If we must make bets about the structure of reality, I think we had better project them from the best representation we have of that reality – and that is modern science as it exists, not as we fancy it might exist.'

are complete in this respect, then, indeed, 'the properties of biology supervene on those of physics'.

However, I am maintaining that the standard physical laws are not complete in this respect. Worse than this, I claim that they are not quite correct in ways that could be importantly relevant to biology. Standard theory allows for an opening of a sort – in the **R**-process of conventional quantum mechanics. On the normal view, this merely gives rise to a genuine randomness, and it is hard to see how a new 'biological' principle could be playing a role here without disturbing the genuineness of this randomness – which would mean changing the physical theory. But I am claiming that things are worse than this. The **R**-procedure of standard theory is *incompatible* with unitary evolution (**U**). Put brutally, the U-evolution process of standard quantum theory is grossly inconsistent with manifest observational fact. In the standard viewpoint one gets around this by means of various devices of differing degrees of plausibility, but the brute fact remains. To my mind, there is no doubt that this is a physics problem, whatever its bearing on biology might be. Possibly it is a coherent viewpoint that a 'patchwork' Nature could simply live with this situation – but I very much doubt that our world is actually like that.

Beyond this kind of thing, I simply don't understand what a biology that does not supervene on physics could be like. The same applies to chemistry. (In this I mean no disrespect to either of those two disciplines.) Some people have expressed to me something analogous in saying that they cannot conceive of a physics whose action is non-computable. This is not an unnatural sentiment, but the 'toy model' universe that I described in Chapter 3 gives some idea of what a non-computable physics could possibly be like. If someone can similarly give me an idea of what a 'biology' that does not supervene on its corresponding 'physics' could be like, then I might begin to take such an idea seriously.

Let me return to what I take to be Nancy Cartwright's main question: why do I believe that we must look to a new physics for a scientific explanation for consciousness? My short answer is that, in accordance with Abner Shimony's discussion, I simply do not see any room for conscious mentality within our present-day physical world-picture – biology and chemistry being part of that world-picture. Moreover, I do not see how we can change biology to be not part of that world-picture without also changing physics. Would one still want to call a world-view 'physics-based' if it contains elements of protomentality at a basic level? This is a matter of terminology, but it is one that I am reasonably happy with for the moment at least.

RESPONSE TO STEPHEN HAWKING

Stephen's comments about his being a positivist might lead one to expect that he, also, would be sympathetic to a 'patchwork' picture of physics. Yet he takes the standard principles of U quantum mechanics to be immutable, as far as I can make out, in his own approach to quantum gravity. I really don't see why he is so unsympathetic to the genuine possibility that unitary evolution might be an approximation to something better. I am, myself, happy with it being an approximation of some kind – as Newton's superbly accurate gravitational theory is an approximation to Einstein's. But that, it seems to me, has very little to do with Platonism/positivism, as such.

I do not agree that environmental decoherence alone can un-superpose Schrödinger's cat. My point about environmental decoherence was that once the environment becomes inextricably entangled with the state of the cat (or with whatever quantum system is under consideration), then it does not seem to make any practical difference which objective reduction scheme one chooses to follow. But without *some* scheme for reduction, even if it is merely some provisional FAPP ('for all practical purposes') scheme, the cat's state would simply remain as a superposition. Perhaps, according to Stephen's 'positivist' stance, he does not really care what the unitarily evolved cat-state actually is, and he would prefer a density matrix description for 'reality'. But this does not, in fact, get us around the cat problem, as I showed in Chapter 2, there being nothing in the density matrix description which asserts that the cat is either dead or alive, and not in some superposition of the two.

With regard to my specific proposal that objective reduction (**OR**) is a quantum gravitational effect, Stephen is certainly correct that 'according to the accepted physical ideas, [space-time] warping will not prevent a Hamiltonian evolution', but the trouble is that without an **OR** process coming in, the separations between the different space-time components can get larger and larger (as with the cat), and seem to deviate more and more from experience. Yes, I do believe that accepted ideas must be wrong at this stage. Moreover, although my ideas are far from being fully detailed as to what I do believe must be going on at this level, I have at least suggested a criterion which is in principle subject to experimental test.

With regard to the likelihood of the relevance of such processes to the brain, I agree that this would seem to be 'very unlikely' – were it not for the fact that something very strange is indeed going on in the conscious brain which appears to me (and also to Abner Shimony) to be beyond what we can understand in terms of our present-day physical world-picture. Of course this is a negative argument, and one must be very cautious not to go overboard with it. I think that it is very important to look into the actual neurophysiology of the brain, and also other aspects of biology, extremely carefully to try to see what is really going on.

Finally, there is my use of the Gödel argument. The whole point of using this kind of discussion is that it is something that *can* be measured from the outside (i.e., I am concerned with the A/C or B/C distinction, as I mentioned earlier, not the externally non-measurable A/B distinction). Moreover, with regard to natural selection, the precise point that I was making was that a specific ability to do mathematics was not what was selected for. If it had been, then we would have been trapped within the Gödelian straight-jacket, which we are not. The whole point of the argument, in this particular regard, is that it was a general ability to *understand* that was selected for – which, as an incidental feature, could also be applied to mathematical understanding. This ability needs to be a non-algorithmic one (because of the Gödelian argument), but it applies to many things other than mathematics. I don't know about earthworms, but I am sure that elephants, dogs, squirrels and many other animals have their good share of it.

APPENDIX 1

Goodstein's Theorem
and Mathematical Thinking

In Chapter 3, I gave a proof of a version of Gödel's theorem, in support of my contention that human understanding must involve ingredients that cannot be simulated by computational procedures. But people often have difficulty appreciating the relevance of Gödel's theorem to the way that we think, even in the case of mathematical thinking. One reason for this is that, according to the way in which the theorem is usually presented, the actual 'unprovable' statement that the Gödel procedure generates seems to have no relevance to any mathematical result of interest.

What Gödel's theorem tells us is that for any (sufficiently extensive) computational 'proof' procedure P, which we are prepared to trust as being unassailably reliable, one can construct a clear-cut arithmetical proposition $G(P)$ whose truth we must also accept as having been unassailably established, but which is inaccessible by the original proof procedure P. The difficulty addressed here is that the actual mathematical statement $G(P)$ which the *direct* application of Gödel's prescriptions provides us with would be enormously difficult to comprehend and of no obvious intrinsic mathematical interest, apart from the fact that we know it to be true but not derivable using P. Accordingly, even mathematicians frequently find themselves happy to disregard mathematical statements like $G(P)$.

Yet, there are examples of Gödel statements that are easily accessible, even to those who have no particular familiarity with mathematical terminology or notation beyond that used in ordinary arithmetic. A particularly striking example came to my attention in 1996, in a lecture by Dan Isaacson (after the Tanner lectures were given, on which this book was based), and I did not know about it when I wrote the material for this book. This is the result known as *Goodstein's theorem*.[1] I believe that it is instructive to give Goodstein's theorem explicitly here, so that the reader can gain some direct experience of a Gödel-type theorem.[2]

To appreciate what Goodstein's theorem asserts, consider any positive whole number, say, 581. First, we express this as a sum of distinct powers of 2:

$$581 = 512 + 64 + 4 + 1 = 2^9 + 2^6 + 2^2 + 2^0.$$

(This is what would be involved in forming the *binary* representation of the number 581, namely, 1001000101, where the 1s represent powers of 2 that are present in the expansion and the 0s represent those which are absent.) It will be noticed that the 'exponents' in this expression, namely, the numbers 9, 6, and 2 could also be represented in this way ($9 = 2^3 + 2^0, 6 = 2^2 + 2^1, 2 = 2^1$), and we get (incorporating $2^0 = 1, 2^1 = 2$)

$$581 = 2^{2^3+1} + 2^{2^2+2} + 2^2 + 1.$$

There is still an exponent at the next order, namely, the 3, for which this representation can be adopted yet again ($3 = 2^1 + 2^0$), and we can obtain

$$581 = 2^{2^{2+1}+1} + 2^{2^2+2} + 2^2 + 1.$$

For larger numbers, we may have to go to third- or higher-order exponents.

We now apply a succession of simple operations to this expression, these *alternating* between

(a) increase the 'base' by 1

and

(b) subtract 1.

The 'base' referred to in (a) is just the number '2' in the preceding expressions, but we can find similar representations for larger bases: 3, 4, 5, 6, and so on. Let us see what happens when we apply (a) to the last expression for 581 above, so that all the 2s become 3s. We get

$$3^{3^{3+1}+1} + 3^{3^{3-1}+3} + 3^3 + 1$$

(which is, in fact, a number of 40 digits, when written out in the normal way, starting 133027946 . . .). Next, we apply (b), to obtain

$$3^{3^{3+1}+1} + 3^{3^2+3} + 3^3$$

(which, of course, is still a number of 40 digits, starting 133027946 . . .). Now apply (a) again, to obtain

$$4^{4^{4+1}+1} + 4^{4^4+4} + 4^4$$

(which is now a number of 618 digits, starting 12926802 . . .).
The operation (b) of subtracting 1 now yields

$$4^{4^{4+1}+1} + 4^{4^4+4} + 3 \times 4^3 + 3 \times 4^2 + 3 \times 4 + 3$$

(where the 3s arise analogously to the 9s that occur in ordinary
base-10 notation when we subtract 1 from 10000 to obtain
9999). The operation (a) then gives us

$$5^{5^{5+1}+1} + 5^{5^5+5} + 3 \times 5^3 + 3 \times 5^2 + 3 \times 5 + 3$$

(which has 10923 digits and starts off 1274 . . .). Note that the
coefficients 3 that appear here are necessarily all less than the
base (now 5) and are unaffected by the increase in the base.
Applying (b) again, we get

$$5^{5^{5+1}+1} + 5^{5^5+5} + 3 \times 5^3 + 3 \times 5^2 + 3 \times 5 + 2,$$

and we are to continue this alternation (a), (b), (a), (b), (a), (b)
as far as we can. The numbers appear to be ever increasing,
and it would be natural to suppose that this continues indefinitely. However, this is not so; for Goodstein's remarkable theorem tells us that no matter what positive whole number we
start with (here 581) we always eventually end up with zero!

This seems extraordinary. But it is, in fact, true, and to get a
feeling for this fact, I recommend that the reader try it – starting first with 3 (where we have $3 = 2^1 + 1$, so our sequence gives
3, 4, 3, 4, 3, 2, 1, 0) – but then, more importantly, trying 4
(where we have $4 = 2^2$, so we get a sequence that starts tamely
enough with 4, 27, 26, 42, 41, 61, 60, 84, . . . , but which
reaches a number with 121210695 digits before decreasing
finally to zero).

What is rather more extraordinary is that Goodstein's theorem is actually a Gödel theorem for the procedure we learn at school called *mathematical induction*.[3] Recall that mathematical induction provides a way of proving that certain mathematical statements S(n) hold for every n = 1, 2, 3, 4, 5, ... The procedure is to show, first, that it holds for n = 1 and then to show that if it holds for n, then it must also hold for n + 1. A familiar example is the statement

$$1 + 2 + 3 + 4 + 5 + \ldots + n = \frac{1}{2}n\,(n + 1).$$

To prove this by mathematical induction, we first establish that it is true for n = 1 (obvious) and then confirm that if the formula works for n, then it also works for n + 1, which is certainly true because we have

$$1 + 2 + 3 + \ldots + n + (n + 1) = \frac{1}{2}n(n + 1) + (n + 1)$$
$$= \frac{1}{2}(n + 1)((n + 1) + 1).$$

What Kirby and Paris demonstrated, in effect, was that if P stands for the procedure of mathematical induction (together with the ordinary arithmetical and logical operations), then we can re-express G(P) in the form of Goodstein's theorem. This tells us that if we believe the procedure of mathematical induction to be trustworthy (which is hardly a doubtful assumption), then we must also believe in the truth of Goodstein's therorem – despite the fact that it is not provable by mathematical induction alone!

The 'unprovability,' in this sense, of Goodstein's theorem certainly does not stop us from seeing that it is in fact true. Our insights enable us to transcend the limited procedures of 'proof' that we had allowed ourselves previously. In fact, the way Goodstein himself proved his theorem was to use an instance of what is called 'transfinite induction.' In the current context, this provides a way of organizing an intuition that can be directly obtained by familiarizing oneself with the 'reason' that Goodstein's theorem is, in fact, true. This intuition can be largely obtained by examining a number of individual cases of Goodstein's theorem. What happens is that the modest little operation (b) relentlessly 'chips away' until the towers of exponents eventually tumble away, one by one, until none is left, even though this takes an incredibly large number of steps.

What all this shows is that the quality of *understanding* is not something that can ever be encapsulated in a specific set of rules. Moreover, understanding is a quality that depends upon our awareness, so whatever it is that is responsible for conscious awareness seems to be coming essentially into play when 'understanding' is present. Thus, our awareness seems to be something that involves elements that cannot be encapsulated in computational rules of any kind; there are, indeed, very strong reasons to believe that our conscious actions are essentially 'non-computational processes.'

There are certainly possible 'loopholes' to this conclusion, and supporters of the computational philosophical standpoint with regard to conscious mentality would have to fall back on one or more of these. Basically, these loopholes are that our capacity for (mathematical) understanding might be the result of some calculational procedure that is unknowable because of its complication, or perhaps knowable in principle but not knowably correct, or it might be inaccurate and only approximately correct. In *Shadows of the Mind,* chapters 2 and 3, I address all these possible loopholes in considerable detail, and I would recommend this discussion to any reader interested in following up these issues more fully. Some readers might find it helpful, first, to examine my *Psyche* account 'Beyond the Doubting of a Shadow.'[4]

NOTES

1. R. L. Goodstein, On the restricted ordinal theorem, *Journal of Symbolic Logic,* **9**, 1944, 33–41.
2. See also R. Penrose, On understanding understanding, *International Studies in the Philosophy of Science,* **11**, 1997, 20.
3. This was shown by L. A. S. Kirby and J. B. Paris in Accessible independence results for Peano arithmetic, *Bulletin of the London Mathematical Society,* **14**, 1982, 285–93.
4. The reference and www address are cited in the footnote on p. 175. The more complete printed reference is: *Psyche* **2**, (1996), 89–129.

APPENDIX 2

Experiments to Test Gravitationally Induced State Reduction

In Chapter 2, I outlined a proposal, according to which a quantum superposition of two states, in which there is a significant mass displacement between the two, should spontaneously reduce – without any outside 'measurement' having to be made on the system – into one state or the other. According to this particular proposal, this *objective state reduction* (OR) is to occur in a time scale of about $T = \hbar/E$, where E is a gravitational energy characterized by the displacement between the two states. For a rigid displacement, we can take this energy E to be the energy that it would cost to displace one instance of the object from the gravitational field of its other instance, this being equivalent to taking E to be the gravitational self-energy of the *difference* between the gravitational fields of the two mass distributions in the two states.

Two developments concerning this issue have taken place in the period that has elapsed since the first publication of this book, one theoretical and the other (proposed) experimental. Both of these have important relevance to Stephen Hawking's complaint (on p. 170) that I have 'not put forward a detailed theory that would enable us to calculate when objective reduction would occur' and to my response (on p. 184) to his remark and to my earlier remark (on p. 90) concerning possible experiments.

On the theoretical side, it has been recognized for some time that there is a certain incompleteness concerning my proposal, as given in this book (on p. 87) and in §6.12 of *Shadows of the Mind* (and a related difficulty with the closely related proposal put forward by Diósi [1989]), where no basic scale parameters are to be introduced beyond the gravitational constant G (and $\hbar$ and c). This incompleteness arises from the fact that there is no clear-cut statement about which states are to be the preferred states that a general state should reduce into. If the preferred states were to be 'position states,' in which each individual particle has a completely well-defined 'pointlike' location, then we should get an infinite value for the relevant gravitational energy E and therefore an instantaneous reduction of any state, in gross violation of many well-confirmed quantum-mechanical effects. But without preferred states of some kind or other, one cannot say which states are to be regarded as unstable 'superpositions' and which are to be taken to be those states (the preferred ones) into which such superpositions are supposed to decay. (Recall that this decay is to have a lifetime $\hbar/E$ according to the OR-scheme. For a finite mass, concentrated at a point, this would lead to $E = \infty$.) In Diósi's [1989] original formulation there is a related problem, namely, energy non-conservation, which, as pointed out by Ghirardi, Grassi and Rimini, would lead to a severe incompatibility with observation. With the introduction of an additional parameter – a fundamental length λ – these authors were able to remove this incompatibility, but there is no a priori justification for choosing any particular value for λ.[1] In effect, in this modified scheme, a state-reduction process would localize some individual particle to a region whose diameter is of the order of λ, rather than to an individual point.

In the scheme that I am proposing, there should be no additional parameter such as λ. Everything should be fixed by the fundamental constants (of relevance) that we already have, namely, G, $\hbar$ and c (and c is itself not of relevance, in the non-relativistic regime). How, then, are the 'preferred states' to be specified? The idea, assuming that velocities are small compared with c and gravitational potentials are also small, is that these states are to be *stationary* solutions of what I call the *Schrödinger-Newton equation*. This equation is simply the (non-relativistic) Schrödinger equation for a wave function Ψ but where there is an additional term given by a Newtonian gravitational potential Φ, where the source for Φ is the expectation value of the mass distribution determined by Ψ. In general, this gives a complicated non-linear coupled system of partial differential equations, and these are still being investigated. Even in the case of a single point particle, it is a non-trivial matter to sort out the stationary solutions of this equation which behave appropriately, including at infinity. But recent work shows that the required solutions do in fact exist for the single point particle, and this gives some mathematical support to the proposal.[2]

The crucial question, of course, is whether a scheme of this nature is in accordance with what actually happens to a macroscopic quantum superposition. It is interesting that certain proposals to test this question experimentally may actually be feasible. Although technically very difficult, these proposed experiments do not seem to require more than that which can be achieved in principle with current technology. The idea is to put a tiny crystal, perhaps not much larger than a speck of dust, into a quantum superposition of two minutely displaced locations, and to ascertain whether this superposition can be maintained coherently for a sizable fraction of a second without its superposed state spontaneously decaying into one state or the other. According to my proposed scheme outlined above, such a decay should take place, whereas the conventional viewpoint would maintain that the superposition would maintain itself indefinitely, unless some other form of decoherence enters to contaminate the state.

Let me describe the general outline of an experimental setup that might be used.[3] The basic experimental arrangement is indicated in the figure. I have illustrated the setup with the incident particle being a photon. However, it should be made clear that this is mainly for ease of description. The ground-based version of this experiment might well be better performed using some other kind of incident particle, such as a neutron or neutral atom of some suitable kind. The reason for this is that the photon that would need to be used in the experiment – if actually a photon – has to be an X-ray photon, and it would be a very considerable technical challenge to construct the required cavity for such a photon. (In the space-based version, the *distance* between the two space platforms that would be used would serve the role of the 'cavity.') For convenience of description in what follows, I shall refer to this particle simply as a 'photon,' whichever form of incident particle is actually intended.

A photon source directs a single photon toward a beam splitter. The beam splitter then effects a splitting of the photon's quantum state into two equal-amplitude parts. One leg of the resulting superposition of the photon's state (the reflected part) is to be kept for, say, about one-tenth of a second without

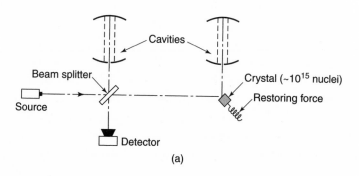

(a)

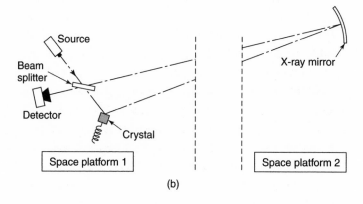

(b)

Fig. 1. (a) Suggested ground-based experiment; (b) suggested space-based experiment.

loss of phase coherence. In the ground-based experiment, this would be achieved by keeping the photon in some kind of cavity; in the space-based experiment the photon is transmitted to an X-ray mirror on a separate space platform, perhaps about an Earth-diameter away. In the other leg of the photon's state, the photon impinges upon a small crystal – containing, say, about 10^{15} nuclei – and the photon is reflected from the crystal, imparting a significant fraction of its momentum to the crystal. In the ground-based experiment, this part of the photon state, reflected from the crystal, is itself kept in a similar (or perhaps the same) cavity as the other part of the photon's state; in the space-based experiment, this second part of the photon's state is also sent to the mirror on the space platform. The crystal is to be such that the entire momentum of the photon's impact upon it is shared by all the crystal's nuclei acting as a rigid body (as with a Mössbauer crystal) without there being a significant probability of exciting internal vibrational modes. The crystal is subject to some kind of restoring force – indicated in the figure by a spring – of such strength that it returns to its original location in, say, one-tenth of a second. At that moment, in the ground-based version, the part of the photon's state which impinged upon the crystal is released from the cavity so that it reverses its path, cancelling the velocity of the returning crystal as it does so. The other part of the photon's state is then also released, with very precise timing, the two parts coming together at the original beam splitter. In the space-based version, the mirror on the space platform reflects each part of the photon's state back to where it (or the other one) came from, on the main space platform, and the result is similar. In either version, provided that there has been no loss of phase coherence in the entire process, the two parts of the photon's state combine coherently at the beam splitter and exit the same way that they came in, so that a detector positioned in the alternative exit beam from the beam splitter would detect nothing.

Now, according to my proposal, the superposition of two crystal locations, which persists for about one-tenth of a second in the preceding descriptions, would be unstable, with a decay time of about that order. This is assuming that the wave function of the crystal is such that the expectation value of the mass distribution of the locations of the nuclei is rather tightly concentrated about their average nuclear positions. Thus, according to this proposal, there would be a large probability that the superposed crystal locations (a 'Schrödinger's cat') will spontaneously reduce in actuality to one location or the other. The photon's state is initially entangled with that of the crystal, so that spontaneous reduction of the crystal's state entails a simultaneous reduction of the photon's state. In this circumstance, the photon has now 'gone one way or the other' and is no longer a superposition of the two, so that phase coherence is lost between the two beams and there is now a significant (calculable) probability that the detector will detect the photon.

Of course, in any actual experiment of this nature, there are likely to be many other forms of decoherence which could destroy the interference between the two returning beams. The idea here is that after all these other forms of decoherence are reduced to a sufficiently small degree, then by varying the parameters involved (size and specific nature of the crystal, the distance it is displaced in relation to lattice spacing, etc.), it would be possible to identify the particular signature of decoherence time inherent in the OR-scheme that I am promoting. There are many modifications of this proposed experiment that can be considered. (In one of these, suggested by Lucien Hardy, two photons are used, and there may be some advantages to the ground-based version in that the individual photons need not themselves be kept coherently for one-tenth of a second.) It seems to me that there is a reasonable prospect of putting to the test, in the not-too-distant future, not only my own OR-scheme, but also various other proposals for quantum state reduction that have been put forward in the literature.

The result of this experiment could have important implications for the foundations of quantum mechanics. There could well be a profound influence on the use of quantum mechanics in many areas of science, such as in biology, where there need be no clear-cut division between 'quantum system' and 'observer.' Most particularly, the suggestions that Stuart Hameroff and I have put forward concerning the physical and biological processes taking place in the brain in order to accommodate the phenomenon of consciousness depend crucially upon the existence and scale of the effects that these experiments are designed to test. A conclusive negative result in these experiments would rule out our proposal.

NOTES

1. Ghirardi, G. C., Grassi, R., and Rimini, A. Continuous-spontaneous-reduction model involving gravity, *Physics Review*, **A42**, 1990, 1057–64.
2. See Moroz, I., Penrose, R., and Tod, K. P. Spherically-symmetric solutions of the Schrödinger-Newton equations, *Classical and Quantum Gravity*, **15**, 1998, 2733–2742; Moroz, I., and Tod, K. P. An analytic approach to the

Schrödinger-Newton equations, to appear in *Nonlinearity*, 1999.

3. I am grateful to a number of colleagues for suggestions in relation to this. Most particularly, Johannes Dapprich suggested the idea that a small (Mössbauer-like) crystal might be the appropriate object to put into a linear superposition of two slightly differing locations. Considerable encouragement about feasibility issues, and specific suggestions about the appropriate scales for the experiment, were put to me by Anton Zeilinger and several members of his experimental group at the Institute of Experimental Physics at the University of Innsbruck. The space-based version of this experiment was the result of discussions with Anders Hansson. See Penrose, R., Quantum computation, entanglement and state reduction, *Philosophical Transactions of the Royal Society of London*, 356, 1998, 1927–39, for a preliminary account of the ground-based version of the experiment.